Python For Beginners

Step-By-Step Guide to Learning Python Programming

Author

Larry Lutz

Table of Contents

Chapter 1

Introduction to Python

Introduction:

The software consists of many small programs interacting with each other. Each program is a combination of instructions in an ordered manner to perform a specific process. Processes are of different types based on the complexity of the program. Development of programs is done based on the type of complexity. Algorithms are needed to find a solution to the problem or implement processes.

An algorithm can be considered as the logic of the program. Each program is written with some type of algorithm. After the development of the program, program testing is done to measure the performance of the program for different inputs. Proper documentation is done with the development of each program for future reference.

To develop any program in the software industry, there are mainly seven stages to follow:

- Requirement Gathering

- Analysing problem

- Decide Input and Output

- Developing Algorithm

- Program implementation

- Testing and Debugging

- Documentation

To find a solution to any software problem, design approaches play a very important role. It is essential to represent the solution for large complex systems. There are various design approaches evolved over the time in the software domain.

Top-Down Approach:

The system consists of various components in a proper hierarchy. In this design approach, designing is done from top-level components to bottom level components.

Bottom Top Approach:

This is the reverse designing approach to Top-down approach. Bottom Level components are designed first and then move to top-level components. Bottom level components are also called base components of the system.

Modular Approach:

This approach is aimed at segregating the whole system into different modules. Each module is implemented differently using a program. Modules are well defined in terms of input and output, it provides flexibility to modify in future and independent testing. Every language is designed on the basis of its requirement and purpose. Like FORTRAN was developed to solve problems related to science and mathematics, COBOL was developed to find solutions related to business applications.

Python interpreter development was started by Guido van Rossum as his hobby project as a successor to ABC Language, but today because of its simplicity and pseudo code characteristics, it has a million users all around the world. Python interpreter is not only able to solve complex programming problems, but able to target problems of the 21st century in the field of automation, web development, desktop application, and many more.

History:

Python was one of the hobby projects for Guido van Rossum after his regular job in the late 1980s. The irrelevant project name was because of his fondness towards Monty Python's Flying Circus. His intention was to develop a simple and readable code interpreter. Guido released the first version of Python interpreter in the year 1991. Today, there are many python versions available in the series names of 2.X to 3.X and still, latest version are releasing every year.

Presently, Python's development and upgradation are handled by a non-profile organization named Python Software Foundation and Guido van Rossum still holds a very important role in the development of python interpreter. There are many versions of python interpreter, and with every release, its feature has been improved and new features included. In October 2000, Python 2.0 was released.

The major features included were Unicode support and Memory Management with a cyclic-detection garbage collection system. In the year 2008, Python 3.0 was released with major functionalities backward compatibility with python 2.6 and python 2.7 version.

Scripting Language:

- Python is a high level and general purpose programming language.
- You might have seen people considering it as a scripting language because they understand script and program as the same.
- They often use the word "Script" instead of "Program." Python has become the tool for many people around the world because of it's easy to use characteristics. Sometimes Python users also infer "python file" by using the term "script."
- Commonly, Python is an Object-oriented programming language that inherits all the advantages of OOP and dividing a program into procedures, modules, and functions. Its object-oriented orientation makes it useful for the scripting purpose.

Advantages:

Python language is widely used all over the globe. Its popularity is because of its characteristics and many advantages attached to it. Some of the major advantages are as follows:

Easy-to-Learn, Read and Maintain:

Python's design philosophy focusses more on the readability of the code. Its pseudo code nature makes it easy-to-learn for beginners who want to learn to programme.

Any non-computer science background can understand by reading the code because of its simple English words used as Keywords. Python's code is also very easy-to-maintain.

A handful of Standard Libraries:

Python's package is available with many standard libraries, which are an aid in solving diverse programming challenges. These libraries are also cross-platform compatible. It allows you to port your Python code to any platform such as Windows, Mac, and Linux.

Easy development and Test:

Python Interactive is very popular and a quick Python interpreter. It helps you to test and run code snippets pretty quick. When you are in the middle of a large program and need to test some code, you just need to run Python interpreter and run into it.

Graphical User Interface programming:

Python avails many libraries for the development of GUI such as Tkinter, Wx, and PyQt, etcetera. These libraries support system calls and cross-platform compatibility.

Extendable to Low-Level languages:

Python also allows you to include low-level programming modules like C, C++, and Java in your code that aid in the development of efficient and fast solutions. Because of its extendable nature, you can have all the advantages of a low-level programming language with quick development.

Disadvantages:

With the numerous advantages of using the Python language over the year in various fields. there are also some downsides of using it for some applications.

o Python is a high-level language, so its execution speed is not as fast as compared to C and C++. But over time, Python libraries are optimized to use it in applications where timing is the important aspect.

o For GUI programming, Python libraries are optimized enough to provide service almost as fast as C and C++.

Exercise

1. Explain the design philosophy of Python.

   Answer: Guido had the following philosophy while designing and implementing:

   - Python's implementation should not be tied to a specific platform. There is no problem if some features are not always available, but the core should work anywhere.

   - Do not disturb the details the machine handles.

   - Mistakes should not be fatal. That is, as long as the virtual machine is still valid, the user code should be able to recover from the error condition.

   - The user's Python code should not be allowed to cause errors in the Python interpreter's undefined behavior; the core dump will never be the user's fault.

2. What are the key features in python?

   Answer: Key features of python is as following –

   - Python is an interpreted language. This means that, unlike C and its variant languages, Python does not need to be compiled before it runs. Other interpreted languages include PHP and Ruby.

   - In Python, a function is a first class object. This means that it is assigned to a variable, returned from another function, and passed to the function. The class is also a first class object.

   - The creation of Python code is fast, but it runs slower than the compiled language.

   - Python is suitable for object-oriented programming. This is to enable class definitions, combinations, and inheritance.

   - Python offers applications in many areas, including web applications, automation, scientific modelling, and large-scale data applications. It is also often used as a "glue" code to make differences in other languages and components.

Chapter 2

Utilities of Python

Introduction:

Based on the various statistics available online, there are almost a million users of the Python language. The numerical data may be more or less than that as Python is open source language and this data is probably based on the number of downloads. Python source code is available online, but Python Software Foundation still holds Copy-rights for this language.

Python's source code is available to use under GNU General Public License. Today, Python package comes pre-installed with Macintosh and Linux operating system. Because of its various impactful characteristics, Python is used in many software solutions and applied to solve real-time problems with profit generating solutions.

Big giants like Google, Netflix, and Dropbox have used the Python language in many ways. The backend process of

Google web search engine is written in Python. The world's largest collection of videos, Youtube, is completely developed in Python. The Dropbox used Python in storage services and for its desktop applications.

Utilities

Besides the well-designed characteristics of Python, Python is used to solve many real-world problems in the various domains. Programmers also use it for solving their day-to-day life problems. In fact, Python applications are nearly unlimited as it can be used from simple gaming applications to high-end complex aerospace and robotics solutions.

Some of the present and emerging applications as described in the following sections:

Graphical User Interface:

Python has a rich set of GUI libraries that could be used developing front-end for applications. These GUIs are supported by Macintosh, Windows, and Linux distributions. Tk library is included automatically with Python 2.0 named Tkinter. This library could also be extended by PMW library to use enhanced widgets in front-end. Qt GUI library is also available with name PyQt and Swing GUI with name Jython. These GUIs are not only available limited to computer applications, but also in embedded applications.

Web-Scripting:

Python has made the complex client-server programming really very simple by the use of standard libraries available with it. These modules let programmers to implement networking task pretty quick. Python scripts also help in creating sockets and data communication over it. File transferring using FTP and parsing XML data is easy-to-implement. There are available methods for network communications for sending, receiving, parsing, and creating e-mails.

Database Programming:

For the demand of accessing the data from the database traditionally, Python also avails features of database accessing and programming for the commonly used databases like MySQL, Oracle, ODBC, and Sybase. It is also considered as the portable database API as it provides the code portability for database just by changing vendor interface.

Mathematical and Scientific Applications:

Python is able to target problems of complex math as well as scientific domain that has not been targeted by any programming language traditionally. NumPy is the very popularly used numeric library, which allows the programmer to solve quick numeric problems in programming application. It is one of the Python's compelling utilities.

There are many more standard libraries available for numeric computations and representation of numeric data in 3-D plot models. SciPy and ScientificPython are popular libraries used as scientific tools that differentiate Python from the other traditional programming languages. These are well optimized in terms of processing the complex algorithms and math. Due to this reason. NumPy is the core interface in the development of SciPy library.

Gaming Application:

Gaming software industries also take advantage of Python libraries such as PyGame, PySoy, Pyglet, and others. Some libraries also include multimedia functionalities with it.

Embedded Applications:

Embedded is a combination of software and hardware component such as microprocessor and microcontroller based applications. Raspberry pi is one such popular microprocessor which uses the Python language for the firmware development. All the modules that control it are written in the Python language.

Image and Data-Mining Applications:

Image processing and Data-mining are the emerging fields in the 21st century. There are various interfaces available that are being used for image processing applications like PyOpenGL, OpenCV, and Maya. Data-mining deals with the large set of data and applying mathematical calculations for generating results, and Python is a great tool for the same. Matplotlib and Mayavi are the common interfaces available modules for data mining and visualization.

Exercise

1. What is the difference between deep and shallow copy?

   Answer:

Shallow copy: When creating a new instance type, use shallow copy and keep the value copied to the new instance. Shallow copies are used to copy reference pointers in the same way as copy values. These references point to the original object, and changes to the members of the class will also affect the original copy. Using shallow copies reduces program execution time and depends on the size of the data used.

Deep copy: Deep copy is used to store the copied values. With deep copy, reference pointers to objects are not copied. It contains a reference to an object and a new object pointed to by another object. Changes made to the original copy do not affect the use of other copies of the object. Deep replication slows program execution because copies of each invoked object are created.

2. Explore more utilities of Python programming language.

   Answer: Many organizations are currently using Python to perform key tasks. Organizations usually have information to publish trade secrets, so you do not necessarily need to listen to these messages. However, Python plays an important role in organizing the way we work and maintain revenue. Here are some of the key ways companies can use Python. This makes Python easier to use in your organization.

   - Corel
   - D-Link
   - Eve – Online
   - Forecast watch
   - Frequentis
   - HP

- Honeywell

Chapter 3

Configuring Python Environment

Before you start with the Python programming, you need Python on your computer. You can check whether Python is already installed on your computer or not. Open your command line windows and type "python" and hit enter, if it displays any response from Python interpreter with the version number then you don't need to download Python on your system.

Python is available on wide variety of platforms. You can download Python for all different environments and it can be ported to the Java and .net virtual machines. For example, you can use python on your UNIX, Linux, Windows, Macintosh, DOS, etcetera.

Getting python:

The most up to date and previous version of Python is available on the official Python website with source code, binaries, and all preferable documentation. You can visit the official Python website at https://www.python.org/.

You can download or refer Python documents from https://www.python.org/doc/. The particular documentation is available for all versions of Python.

Installing Python:

Python is available for wide variety of platforms. You need to download the binary file of Python version according to the platform and then install Python on your computer.

If the binary code is not available for your platform, then you can use a C compiler to compile the source code manually. Compilation of source code gives more flexibility in terms of choice of features.

Windows Installation:

Python interpreter is not pre-installed in Windows, but it does not mean that Windows users won't find a useful, flexible programming language. However, installing the latest version of Python is not a trivial matter, so you make sure to find the right tool for the task.

You can download latest version of Python 2 and Python 3 according to your need. The installer will install 32-bit or 64-bit version according to your computer automatically.

Python 2 Installation:

You can install Python 2 version from the official Python site https://www.python.org/downloads/. The latest version is also available but if you want to download an older version then you can do it by downloading its binary code. Click on Download Python 2.7.14 so it will start downloading binary code on your computer automatically.

- While downloading, the installer will set a path variable for you. Download and Run the installer.

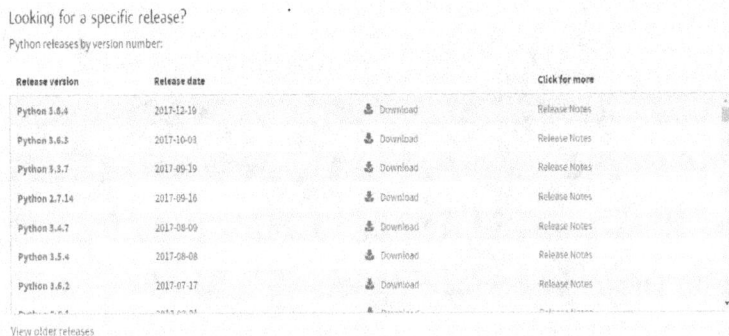

- Select Install for all users and click on Next button.

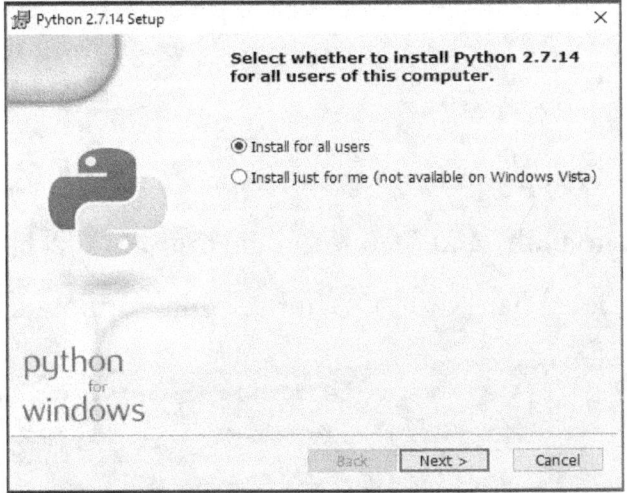

- While selecting the directory, leave the directory as Python 27 and click on Next button.

- On customize Python screen, click on "Add python.exe to path" and then select the option "Will be installed on local hard drive." After selecting option, click on Next button.

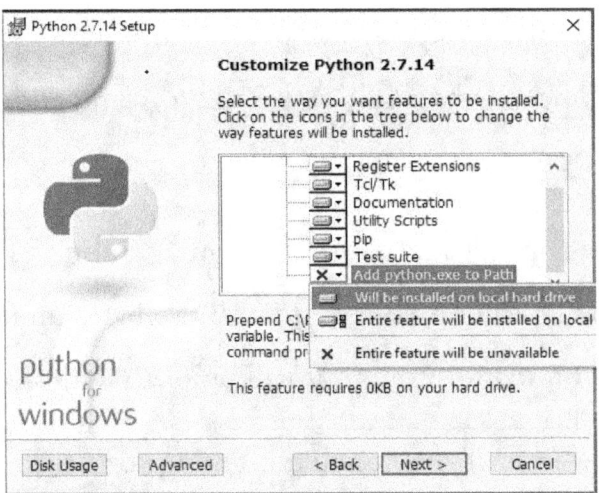

- After clicking on Next, it will start the installation process. After completing it, click on Finish button.

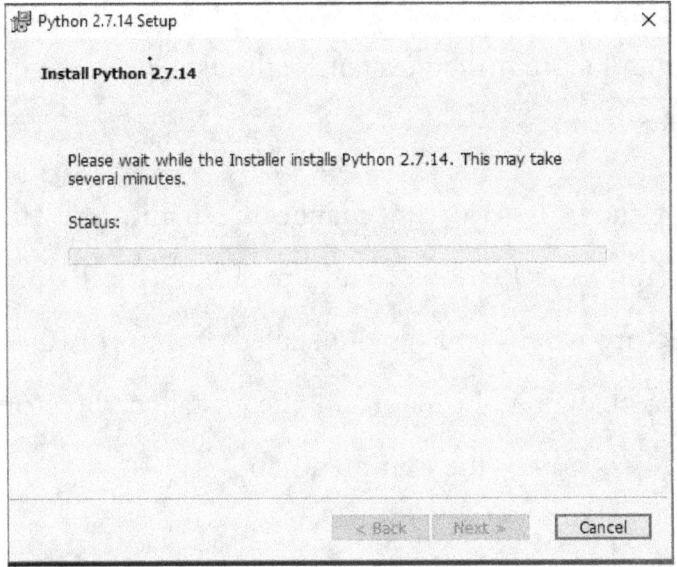

- You can search for python in "Start" Menu and can open Python Idle and Python Command line for more operation.

You can install and work on both Python version 2 and 3 simultaneously on your system. But when you type "Python" at the command prompt, it will point to Python 2.7.

It is because of the variable pointing to the directory and all executable present in that directory that works as the command in the command line. If two directories are present and both have a "python.exe" file, then it will use a variable of the directory which is higher in the list. If there is variable for user and system then system path takes precedence over user path.

To remove it, you can change the name of Python folder, "python" for Python 2 and "python3" for Python 3 in the directory where you have installed Python on the computer. After changing the name in the installed directory you can check version in the command line.

If you are not satisfied with this solution, then you can reorder the environment variable and use Python version according to the need of your project.

Linux Installation:

You can install and setup Python by using Terminal, which is non-graphical. Instead of selecting options from GUI screen and click on buttons, you need to write commands and receive feedback from your computer.

Ubuntu 16.04 comes with the Python 2 and Python 3 pre-installed. To make sure that you have the latest version of Python or not, you can update and upgrade your computer with apt-get command.

$ sudo apt-get update

$ sudo apt-get –y upgrade

The -y flag will confirm that you are installing all the projects in the system, but depending on your Linux version, you need to select additional prompts during system updates and upgrades.

If you are using an older version of Ubuntu and other Linux based operating systems in which Python is not pre-installed, then you can use the following command for installing Python:

$ sudo apt-get install python2.7

You can check the version of installed Python by typing the command:

$ python2.7 –V

You will receive output with the Python version in the terminal window. The output will look like this:

Python 2.7.14

To install pip in your system, use command:

$ sudo apt-get install –y python-pip

Macintosh Installation:

The installation process of Python is somewhat similar in the Linux and Macintosh. Macintosh comes with the pre-installed Python version. You can check the version of Python by typing:

$ python –V or

$ python --version

If you are installing Python again or need to install the latest version of Python, then you need to type the following command:

$brew install python

Python IDEs:

There are many Python IDE (Integrated Development Environment) that are useful for you to work on a Python project. IDE can easily handle big projects that have hundreds of small modules. The main focus while working on IDE is on simplicity and ease of use. It provides a graphical interface to the programmer for their ease. Some of them are very lightweight and fast while working on Python projects. Here is the list of IDEs which are compatible with Windows, Linux, and Mac:

- IDLE
- PyDev
- Eric
- LiClipse
- NetBeans
- Pycharm
- Pyscripter
- Spyder
- Python tools for Visual Studio

There are some of the IDEs which comes with the integrated GUI builder. It is useful when you are working on any Python based GUI projects. They are:

- MonkeyStudio
- Xcode
- Visual Python
- PythonCard

Exercise

1. Install the Python package in your system.

   Refer to the steps as your operating system specified in the chapter.

Chapter 4

Basics of Python

In the previous chapter, we have learned about the various characteristics, utilities, and configuration of Python in your systems. Now it's the time to start with the understanding of the basic Python scripts and get familiar with the programming environment of it.

We will be using the Linux distribution environment for writing and running the scripts throughout this book. You can use any environment other than Linux such as Macintosh or Windows and configure your Python package as explained in the previous chapter and get set go.

Hello, World Program

Hello, World is the most basic program to learn any computer programming language with. Python is a really quick and dirty programming language. It's just a matter of time that you have any solution in your mind and you quickly code it, thanks to the pseudo-code design philosophy of python.

You can either use Python Interactive Mode to write your first Hello, World program and it will provide a prompt output or by using the traditional method for writing program in a text editor. Here both ways have been explained:

Let's get started,

By using Text Editor:

Open any of your favourite text editor such as Notepad, VI editor, or VIM etcetera. Write the following program in it and save it with the name "helloWorld.py".

$ print "Hello, World"

```
# Hello World Program
print "Hello, World!!"
```

To run the program, you need to open Terminal (in Linux) or Command line window (in Windows). Change your current directory to the program file folder and run following command:

$ python helloWorld.py

The output is as follows:

```
Hello, World!!
```

By using Python Interactive:

Python Interactive is a mode of python which programmers use to write a quick code snippet and test it while working with the large programs. The significance and utility of Python Interactive is explained more in-depth in coming section of this chapter.

Here you can just have a hands-on experience of Python Interactive with this example.

Let's start with it. Open Terminal in your system and Type following:

$ python

```
Python 2.7.12 (default, Nov 19 2016, 06:48:10)
[GCC 5.4.0 20160609] on linux2
Type "help", "copyright", "credits" or "license" f
>>> █
```

With this command, you are calling Python interpreter to run in interactive mode. Write following one line Hello, World program and then press Enter.

% >>> print "Hello, World"

With the Enter button, promptly it will display output as following:

```
>>> print "Hello, World"
Hello, World
```

This method of running your Hello, World program is just for your experience with Python Interactive. To write any further programs, we will be using traditional text editor method.

Let's understand more in-detail about the steps and the program that we have run.

To get a brief of the one-liner program. Break it into two parts: one is print and the other one is "Hello, World." Print is a function call, similar to the printf function in C. It displays data to output screen in string form and "Hello, World" is a string which is an argument to print function. As of now, you should not think too much about strings and function. These are explained well in the following chapters.

Python Interactive:

The interactive window of Python is simple and also very useful for the programmer during the development of Python code. It is similar to sitting in-front of Python interpreter and getting results for each Python expression. This aids programmers in experimenting and testing of code snippets.

To wake up Python interactive mode in your system, you just need to type Python and hit enter in Terminal. It will next display a few lines with details of python interpreter like version number and others and prompt your for input with ">>>" As showing in the following:

```
Python 2.7.12 (default, Nov 19 2016, 06:48:10)
[GCC 5.4.0 20160609] on linux2
Type "help", "copyright", "credits" or "license" for more information.
>>> █
```

When you work with Python interactively, it will give the result of each expression in the next line as you type it and press Enter key. Due to this, it is not required for you to put a print command in Python interactive mode. Here in the following expression: x = 10, which represents x is assigned with integer value 10 and on pressing Enter key, its value is being displayed. Similarly for str = "Hello, World". The string "Hello, World" is assigned to str variable.

```
>>> str = "Hello, World"
>>> str
'Hello, World'
```

Now It is certainly clear the reason behind the use of Python interactive mode. Being a smart programmer, you can experiment with a few lines of Python commands to see the behaviour of Python when working with large programs.

Basic Built-in Function:

Python package is available with several useful input-output functions. Being a beginner Python programmer, it is a must to understand and remember these functions. These functions' names, syntax, and descriptions are as following:

1. raw_input() or input() :
   This function is similar to the scanf function in C. It is used to take input from the user.
2. print():

This function is useful for printing the data to the output windows in string form.

3. len():

    This function is used to get the length of the object. Here the object can be a string, a tuple, or a list and the object is passed as an argument in the len function.

4. str():

    This function is useful for converting the type of object. Object version is changed to string type.

5. abs()

    This is a mathematical function and it is same as the absolute maths function. It provides the absolute value of the object.

6. help()

    This function is very useful for getting information of any function, method or keyword. If no object is passed in the function, It will prompt to a Python help window, and if any string is passed through it as an object, then it will search for that string in the documentation and shows relevant function or data.

7. min():

> This function gives the smallest element in an iterative object or it will give the smallest element when multiple objects are passed.

8. max()

> this function gives the largest element in an iterative object or it will give the largest element when multiple objects are passed.

9. all()

> This function returns a Boolean value that is either True or False. It gives True as the return value when all the elements in the iterative object elements are true.

10. any():

> This function also returns a Boolean value. It gives True as a return value when any of the elements in the iterative object elements are true.

Exercise

1. Create and Run a program to display following string text.

   "python is widely used programming language"

   Code:

```
print "python is widely used programming language";
```

   Output:

```
python is widely used programming language
```

2. Create a python program to take input string from the user and display it on output window.

   Code:

```
str = raw_input("Enter string: ");
print "input is: ", str
```

Output:

```
Enter string: Hello python
input is:  Hello python
```

Chapter 5

Variables, String and Operators

Variables are the identifier which reserve location in the memory to store values. It means when you are creating any variable, it is creating some space in the memory.

The interpreter will allocate memory based on the data type of variable, and data type defines the type of value the variable holds. The variables can hold integer, character, string, and other data types

Variables(Values):

A value is a small unit of the program like letter and number, which is used while assigning to the variable. We don't need to declare a variable before assigning value. Python interpreter will automatically assign the type of data while assigning the value to that variable.

The = sign is used for assignment. The left part of the equal sign is a variable and right part of the equal sign is a value which is assigned to that variable.

Code:

```
name = "Mark"
height =6
age = 25

print (name)
print (height)
print (age)
```

Output:

```
Mark
6
25
```

In the above code, the variables are "name", "age", and "height" and we are assigning the values to each variable. The variable name is storing the character values, age is storing integer value, and height is storing the float value. We don't need to declare the data type of variable; it will automatically assign data type according to the assigned values.

Rules for variable name:

- The variable name must start with an underscore or character.
- The variable name is case-sensitive and contains only alphanumeric character.
- The variable name can't contain any spaces.
- You can't use reserved words as a variable name.

Data Types:

A variable can hold different types of data in the memory. For storing a name, a string is used, age in numeric value, height in float value. There are some standard data types in Python programming language that you can use for storing data in the memory.

These are the standard data types are:

- String
- Tuple
- Dictionary
- Numbers

- List

Strings:

In the Python language, a string is a sequence of text and bytes. A string starts with a single and double quote. You can also use single quotes within double quotes and vice versa.

In simple words, a string is an array of characters and you can use indexing to access the elements of an array. The index starts at 0 on the left and -1 on the right. In Python, strings are immutable in nature. You cannot change character in string once it is generated. The 'in' operator is used when we need to check presence of substring in the string. The result of matching the string is represented in the form of Boolean value.

Python provides us the very simple method to cut the substring from a string. It is known as string slicing. You can separate two indices by the colon (:).

How to access string Values?

Python language does not support character datatype because the character is treated as a string in Python. It gives a length of string and hence it is considered as a substring.

Code:

```
char1 = 'Hello Python'
str1 = "Python Programming"
print ("First value is: " , char1)
print ("Second value is: ", str1)
```

Output:

```
('First value is: ', 'Hello Python')
('Second value is: ', 'Python Programming')
```

Update String:

Reassigning an existing string with new string will give you updated string. The new string can be related to the previous string or completely new string.

Code:

```
char = "Hello Python"
print ("New String is: " , char)
```

Output:

```
('New String is: ', 'Hello Python')
```

Escape Character:

Backslash Notation	Description
\a	Alert
\b	Backspace
\cx	Control X
\e	Escape
\f	Form feed
\n	New line
\r	Carriage return
\s	Space
\t	Tab
\v	Vertical Tab

Tuples:

A tuple is another type of data type which consists of series of comma- separated values. Like strings, tuples are also immutable and enclosed in the parenthesis with holding mix data type. Like strings, tuples can also be sliced. When we slice tuple, it will create a new tuple, but it does not change the original tuple. Addition(+) Operator is used to create a new tuple that is concatenation of more than two tuples. We use * operator to repeat a tuple.

Code:

```
tuple = ('python', 465, 'language', 70.8)
tuple1 = (458, 'program')

print tuple
print tuple[1]
print tuple[1:2]
print tuple[2: ]
print tuple1 * 2
print tuple + tuple1
```

Output:

```
('python', 465, 'language', 70.8)
465
(465,)
('language', 70.8)
(458, 'program', 458, 'program')
('python', 465, 'language', 70.8, 458, 'program')
```

Dictionary:

In the Python language, dictionary data type is like a hash table. It works like an associative array and hashes similar to Perl. Basically, it consists of key-value pairs. A dictionary key is generally a number and a string but it can be of any Python data type. The values can be like arbitrary Python object.

Code:

```
dictionary = {}
dictionary['one'] = "This is one"
dictionary[2] = "This is two"
dictionary1 = {'name': 'Mark', 'EID' : 4578, 'dept' : 'marketing'}
print dictionary['one']          # Print values for 'one' key
print dictionary[2]              # Print value for key 2
print dictionary1                # Print complete dictionary
print dictionary.keys()          # Print all key
print dictionary.values()        # Print all values
```

Output:

```
This is one
This is two
{'dept': 'marketing', 'name': 'Mark', 'EID': 4578}
[2, 'one']
['This is two', 'This is one']
```

Numbers:

The Number data type is used to store numerical values like 1, 2, etcetera. It is used when programmers need to assign a numeric value to the variable. For example,

age = 25

60

height = 6

Del is used when you want to delete a single or multiple objects. For example,

del age

del age, height

Generally, there are four types of numeric value that you can use in python :

- int (signed integer)
- long (it can be represented in octal and hexadecimal)
- float (floating point values)
- complex

Basic Operator:

The operators are symbols which are used to perform mathematical and logical operations. Operands are the values on which the operator is applied while operations.

Types of Operators:

- Assignment operator

- Logical operator

- Arithmetic operator

- Relational operator

- Bitwise operator

- Identify operator

- Membership operator

Arithmetic Operator:

Symbol	Operator Name
+	Addition
-	Subtraction
*	Multiplication
/	Division
%	Modulus
**	Exponent
//	Floor Division

Logical Operator:

Symbol	Operator Name
or	Logical OR
and	Logical AND
Not	Logical NOT

Assignment Operator:

Symbol	Operator Name
=	Equal
+=	Add AND
-+	Subtract AND
*=	Multiply AND
/=	Division AND
%=	Modulus AND
**=	Exponent AND
//=	Floor Division AND

Relational Operator:

Symbol	Operator Name
==	Double Equal
!= or <>	Not Equal To
>	Greater Than
<	Less Than
<=	Less Than Equal To
>=	Greater Than Equal To

Bitwise Operator:

Symbol	Operator Name
&	Binary AND
|	Binary OR
^	Binary XOR

~	Binary 1s Complement
<<	Binary Left Shift
>>	Binary Right Shift

Identity Operator:

Symbol	Operator Name
Is	Is
Is not	Is not

Membership operator:

Symbol	Operator Name
In	In
Not in	Not in

Exercise

1. Explain Variable and write a code using it.

Answer: Variables are the identifier which reserve memory location to store values. It means when you are creating any variable, it is creating some space in the memory.

Code:

```
name = "Mark"
height =6
age = 25

print (name)
print (height)
print (age)
```

Output:

```
Mark
6
25
```

2. Explain Strings and write a code using it.
   Answer: A string is a sequence of text and bytes. A string starts with a single and double quote. You can

also use single quotes within double quotes and vice versa.

Code:

```
char1 = 'Hello Python'
str1 = "Python Programming"
print ("First value is: " , char1)
print ("Second value is: ", str1)
```

Output:

```
('First value is: ', 'Hello Python')
('Second value is: ', 'Python Programming')
```

3. Explain Operators and name different types of Operators?

Answer: The operators are symbols which are used to perform mathematical and logical operations. Operands are the values on which the operator is applied while in operations.

Types of Operators:

- Assignment operator
- Logical operator
- Arithmetic operator
- Relational operator
- Bitwise operator

- Identify operator

- Membership operator

Chapter 6

Mathematical Aspects

Introduction:

Mathematics is one of the integral parts of programming. Be it a simple maths operation or writing a complex mathematical algorithm for software, python is always ahead in terms of its speed and quick coding practices. Mathematical data is taken as the data object in the Python language. In fact, objects are the building block of Python programming. We will be learning the usage of basic mathematical functions which are frequently used during Python programming. There are many popular and optimized mathematics and scientific libraries available, which are either built-in or can be imported into your Python code to use.

Basic Mathematics Operations:

In addition to the simple operation like addition (+), subtraction (-), multiplication (*), and division(/), there are many mathematical functions available in Python. As the new versions of Python releases, more mathematical functions are added to the package. In Python version 2.7, there are many methods available in the math library such as numeric theoretical functions, power and logarithmic functions, trigonometric and hyperbolic function, and some special functions.

Code:

```
"""
Program to peform basic arithmatic operation
"""
num1 = int(raw_input("Enter Input 1 :"))
num2 = int(raw_input("Enter Input 2 :"))

print "Addition is %d" % (num1 + num2)
print "Subtraction is %d" % (num1 - num2)
print "Multiplication is %d" % (num1*num2)
print "Division is %d" % (num1/num2)
```

Output:

```
Enter Input 1 :45
Enter Input 2 :78
Addition is 123
Subtraction is -33
Multiplication is 3510
Division is 0
```

In the following sections, we will look more into its usage and prototype:

1. Numeric-theoretic Functions:

This module already comes with a built-in Python version 2.7. It is similar to C math library. These functions take one or two objects as data, but it does not take any complex number as objects. Cmath is another Python librarycmath, which is available for complex number math operations.

math.ceil(a) : This function is the same as the ceil function in mathematics. It provides ceil value of 'a' with float datatype. It is the smallest integer value which is greater than or equal to 'a'.

math.floor(a): This function is same as the floor function in mathematics. It provides floor value of 'a' with float datatype. It is the largest integer value which is less than or equal to 'a'.

math.factorial(a) : It returns factorial value of 'a', where 'a' is a positive and integer data, otherwise, it throws an error.

math.fabs(a) : This function returns absolute value of 'a'.

math.copysign(a, b): This function is used to change the sign of the number. It returns data of 'a' with the sign of 'b'.

Let's understand above functions further with Python programming. In the below program:

Code:

```
"""
Program to perform math function
"""
import math

a = 1.456

print "Ceil value of a is %d" % math.ceil(a)
print "Floor value of a is %d" % math.floor(a)
print "Absolute value of a is %d" % math.fabs(a)

b = 5
c = -5
print "Factorial of b is %d" % math.factorial(b)
print "Copied Sign value of b is %d" % math.copysign( b, c)
```

Output:

```
Ceil value of a is 2
Floor value of a is 1
Absolute value of a is 1
Factorial of b is 120
Copied Sign value of b is -5
```

In addition to the above function, a few more functions are available in Python. You can have a look at the python documentation for more.

2. Power and Logarithmic Functions:

Python provides following functions in this category:

math.pow(a,b): This function returns 'a' raised to the power of 'b, where both the data objects should have valid data. For invalid data, the function throws an error.

math.sqrt(a): This function returns square root value of 'a'.

math.log10(a): This function returns logarithmic value of 'a', where the base of the logarithm is 10.

math.log1p(a): This function returns natural logarithmic value of 'a', where base of logarithm is e (constant)

math.exp(a): This function returns exponential value of 'a'

Code:

```
"""
Program to peform math functions.
"""
import math

a = 10
b = 2

print "Power of a raised to b :%d" % math.pow( a, b)
print "Square root of a :%d" % math.sqrt(a)
print "Logarithmic Value of a(base-10) :%d" % math.log10(a)
print "Logarithmic Value of a(base-e)  :%d" % math.log1p(a)
print "Exponential of a :%d" % math.exp(a)
```

Output:

```
Power of a raised to b :100
Square root of a :3
Logarithmic Value of a(base-10) :1
Logarithmic Value of a(base-e)  :2
Exponential of a :22026
```

3. Trigonometric and Hyperbolic Functions:

Python has all trigonometric and hyperbolic functions available in its package. These functions' returns value in radian unit and functions same as the mathematics trigonometric and hyperbolic function.

Trigonometric:

math.sin(a)

math.cos(a)

math.tan(a)

Hyperbolic:

math.sinh(a)

math.cosh(a)

math.tan(a)

Code:

```
"""
Program to perform trigonimetric function
"""
import math

a = 10

print "Sine of a :%f" % math.sin(a)
print "Cosine of a :%f" % math.cos(a)
print "Tan of a :%f" % math.tan(a)
print "Hyperbolic Sine of a :%f" % math.sinh(a)
print "Hyperbolic Cosine of a :%f" % math.cosh(a)
print "Hyperbolic Tan of a :%f" % math.tanh(a)
```

Output:

```
Sine of a :-0.544021
Cosine of a :-0.839072
Tan of a :0.648361
Hyperbolic Sine of a :11013.232875
Hyperbolic Cosine of a :11013.232920
Hyperbolic Tan of a :1.000000
```

4. Special Functions:

Other than the standard maths functions, Python also provides special mathematical functions. They are as following:

math.gamma(a):

This function returns mathematics gamma function value of 'a'.

math.lgamma(a):

This function is a combination of natural logarithm and gamma function. First, it finds the gamma function value at 'a', and then returns natural logarithmic value of absolute value of the result.

math.erf(a):

This function returns error function value at 'a'.

math.erfc(a):

This function returns complementary error function value at 'a'.

Till now, we have discussed the basic mathematics function. Python is also rich with its advanced mathematical capabilities. Its richness also attracted people from research and scientific backgrounds. NumPy, SciPy, and Matplotlib are very well contained and optimized libraries. Every Python programmer must be well versed with these libraries to enhance their Python programming skills. We will learn more in-depth about these libraries and its utilities.

NumPy Library :

NumPy is a short name for Numeric or Numerical Python and developed as the open source project by Travis Oliphant. The key idea behind the development of this library was to handle multi-dimensional data (array) in Python. It was developed by merging two predecessor libraries, one is Numeric and another is Numarray.

NumPy has power to process multi-dimensional array at fast speed. There are following operations you can perform using NumPy.

1. Logical and mathematical operations on multi-dimensional data or matrix.
2. Linear algebra and generating random numbers.
3. Fourier transforms.

Usually, this library won't be pre-installed with your Python package. You need to install it separately using Pip Python module using the following command in Terminal.

$ pip install numpy

There are following methods and functions which are available in NumPy library.

1. Numpy.zeros(a,b,c)

The function creates a new array with all elements entries as zero. Where

'a' is the shape of new array or size of array.

'b' is the datatype for the elements and, it is optional.

'c' is the order of array and it is also optional.

2. Numpy.ones(a,b,c)

   This function creates a new array with all elements entries as one and data objects are same as of zeros function.

3. Numpy.full(a,b,c,d)

   This function returns a newly created array and provides shape and value where

   'a' is the shape of new array or size of array.

   'b' is the value to be fill in the array.

   'c' is the datatype for the elements, and it is optional.

   'd' is the order of the array, and it is also optional.

   Code:

```
"""
Program to perform numpy function
"""
import numpy

# 1-d array of zeros
arr1 = numpy.zeros(5)
print "arr1 :"
print arr1

# 2-d array of zeros
arr2 = numpy.zeros((3,2))
print "arr2 :"
print arr2


# 1-d array of ones
arr3 =  numpy.ones(3)
print "arr3 :"
print arr3

# 2-d array of ones
arr4 =  numpy.ones((2,3))
print "arr4 :"
print arr4

# array of any scalar value
arr5 = numpy.full(5, 10)
print "arr5 :"
print arr5
```

Output:

```
arr1 :
[0. 0. 0. 0. 0.]
arr2 :
[[0. 0.]
 [0. 0.]
 [0. 0.]]
arr3 :
[1. 1. 1.]
arr4 :
[[1. 1. 1.]
 [1. 1. 1.]]
arr5 :
[10 10 10 10 10]
```

SciPy Library :

SciPy name stands for Scientific Python. It is an extension of Python NumPy library to enhance its processing and algorithmic capabilities. As NumPy provides methods for creating multi-dimensional data and its processing in Python, SciPy is one step ahead. It is specifically built for implementation of scientific processing like writing mathematical algorithms application. Because of it, Python is a perfect language if you are programming for niche applications such as scientific, web, and desktop applications.

For installing SciPy in your system, you need following commands in your Terminal window.

$ sudo apt -get install python-scipy

SciPy library is structured into various sub-packages and each sub-package is specific to particular computing domain. These sub-packages and their computing domains are as following :

1. constants: Mathematical Constants.

2. Fftpack: Fast Fourier transform functions.

3. Interpolate: Interpolation functions.

4. Cluster: Clustering algorithms functions.

5. Io: Input and Output.

Before using any sub-package function in your program. You need to import library, for example:

$ from SciPy import constants, io

It is time to go deep into the SciPy library, We will understand some basic functions one-by-one and quickly program it. Let's get started:

1. Constants: The SciPy contains various constant values which are used in both scientific and mathematical calculations. Constants like c (Speed of Light), h (Plank's constant), e (elementary Charge), etcetera.

2. Fftpack: In signalling related applications Fftpack is vastly used. There are many transforming functions present in it.

fft(x[, n, axis, overwrite_x): It is used for generating discrete Fourier transform of any real or complex sequence.

Ifft(x[, n, axis, overwrite_x): It is used for generating discrete inverse Fourier transform of any real or complex sequence.

fft2(x[, shapes, axis, overwrite_x): It is used for finding 2-dimensional Fourier transform.

Ifft2(x[, shapes, axis, overwrite_x): It is used for finding 2-dimensional inverse Fourier transform.

3. Interpolate: In this domain, functions related to various mathematical interpolation methods are available. These functions and their descriptions are as follows:

Interp1d(x,y[,kind,axis,copy,...]): It is used for interpolation of one-dimensional function.

KroghInterpolator(xi,yi[,axis]): It is used for interpolation of a set of points.

4. Cluster: Clustering is one of the latest methods you can use in information theory, compression of data, and detection of the targets. Further, the cluster subpackage contains two modules. One is vq and another is hierarchy. Vq is particularly used for vector quantisation and K-mean algorithm. Hierarchy module supports hierarchical clustering.

5. Io: SciPy supports reading from files and writing to files in various formats. It could be any data like text, numeric, or binary. You can use file like Matrix Market file, Matlab file, IDL files, etcetera.

Code:

```
"""
Program to perform scipy function
"""
import scipy, numpy
from scipy import interpolate

# Constant values
print "Value of e :" + str(scipy.e)

# fftpack
y  = scipy.fft([1.0, 2.0, 3.0, 1.5])
print "Fast Fourier Transform of y :" + str(y)
```

Output:

```
Value of e :2.71828182846
Fast Fourier Transform of y :[ 7.5+0.j  -2. -0.5j  0.5+0.j  -2. +0.5j]
```

Matplotlib Library:

With the enhanced capabilities of Python using NumPy and SciPy. Matplotlib is one of the alternatives of MATLAB software for representation of data and its analysis. With open-source nature Python, these libraries are well used among data scientist and researcher.

You can use Matplotlib for plotting 2-dimensional and 3-dimensional data. It also includes error charts, histogram and bar charts in just a few lines of codes. It makes hard and complex data analysis very easy.

Exercise

1. Perform the following mathematical equation:

   $a((a + b)/(a\text{-}b)) + b + 1$

   where a = 10 and b = 5

   Code:

   ```
   a = 10
   b = 5

   c = a*((a+b)/(a-b)) + b + 1

   print "Output Value is : %d" %c
   ```

   Output:

   ```
   Output Value is : 36
   ```

2. Take an input array from the user and find its Fast-Fourier transformation.

   Code:

```
import scipy

# input from user
print "Enter an 1-d array :"
inputArray = [int(x) for x in raw_input().split()]

# performing fase-fourier transform on input
output = scipy.fft(inputArray)

# displaying output
print "Fast-Fourier Transform is :" + str(output)
```

Output:

```
Enter an 1-d array :
1 2 1 4 1 5
Fast-Fourier Transform is :[14. +0.00000000e+00j -0.5+2.59807621e+00j  0.5+2.598
07621e+00j
 -8. +5.77315973e-15j  0.5-2.59807621e+00j -0.5-2.59807621e+00j]
```

3. Plot the sine wave using Python program.

Code:

```
"""
Plotting Sine wave
"""
import numpy as np
import matplotlib.pyplot as plot

# range of sine wave
time = np.arange(0, 10, 0.1)

# finding amplitude of sine wave
amp = np.sine(time)

# plotting the sine wave
plot.plot(time, amp)
plot.show()
```

Output:

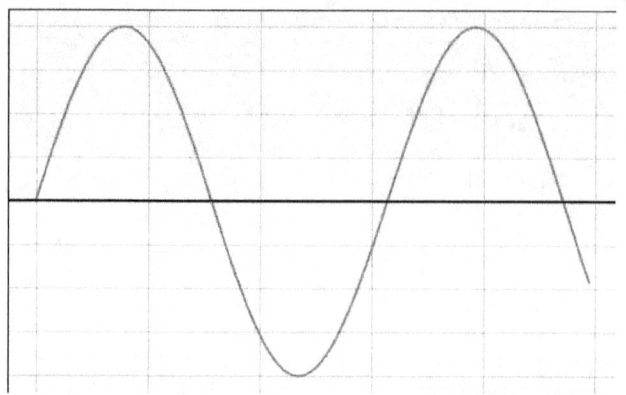

Chapter 7

Data types

In Python, data takes the objects of different types of form – they are either built-in objects provided by the Python language or created by the programmer using Python classes or external language tools. Objects are just pieces of memory used for storing values and set operations on that variable.

Importance of Built-in Types:

In the lower-level language such as C and C++, most of programmer's effort goes into implementing objects to represent the component in the application domain. Being a good programmer, you need to manage memory allocation, memory structure, implement search, and access routine. These chores are very tedious and always distract you from your programming goal.

In the Python language, most of the work goes away as you don't need to do object implementation before you start solving problems. It is always best way to use built-in object instead of implementing your own object.

- Easy to write a program with the help of Built-in Object: Built-in object gives you a collection of lists and dictionaries for free, which is very helpful while working on any task.

- For the complex problem, you need to write your own object with the Python classes and C language interfaces, but it is easier to use built-in types such as list and dictionaries.

- Built-in objects are more efficient than custom data structure because that is already optimized similarly to all data structure algorithms, which is used in C programming language.

To some readers, object types are more powerful while programming. Especially Lists and Dictionary are the more powerful data types, which are very useful in collections and searching in lower-level programming language. Lists provide an ordered collection of objects while dictionary stores object by keys. Lists and dictionary are nested in nature and can grow and shrink according to the demand and also capable to containing the object of any type.

Built-in Objects:

Object Type	Examples
Number	146,2.75,3+4j
String	'python', "programming"
Lists	[1,[4,'six'],6.8]
Dictionaries	{'python': 'programming'}, dict(day=10)
Tuples	(1,'python','U')
Sets	Set('xyz')
Files	Open('egg.txt')
Program unit types	Function, modules, and classes
Implementation relates types	Complied code, stack tracebacks

Other core types	Boolean, Types

Numbers:

Python object also includes the numbers and it contains integer, floating-point numbers, complex numbers, decimal, rational numbers. It supports mathematical operation. For example, plus sign (+) is used for addition, star (*) is used for multiplication, and two stars (**) are sued for exponentiation operation.

Code:

```
>>> 146 + 485
631
>>> 1.3 * 9
11.700000000000001
>>> 2*16
32
>>> 
```

Besides the expression, you can also use numeric modules, which are shipped with Python-modules. They are just Python packages that you can import and add to your program for ease.

Code:

```
>>> import math
>>> math.sqrt(25)
5.0
>>> import random
>>> random.choice([1,2,3,4,5,6])
1
>>>
```

Strings:

Strings are usually used to represent both textual and arbitrary information. String supports an operation that includes positional ordering among items. For example, if you want to calculate the length of a string which is inside quotes, you can use built-in len function and calculate its length.

```
>>> Var = 'Python'
>>> len(Var)
6
>>> Var[2]
't'
>>>
```

In Python, indexes are coded as offset and it start from 0: the first item which is at first place is index 0, second is index 1, and so on. In Python, you can also use of index backward from the endpoint. Positive indexes are counted from the left-hand side and negative numbers count from the right-hand side.

In simple positional indexing, sequences also support slicing in, which you can extract entire section from the string in a single step.

```
>>> Var = 'Python'
>>> len(Var)
6
>>> Var[-1]
'n'
>>> Var[-2]
'o'
>>> Var
'Python'
>>> Var[1:3]
'yt'
>>>
```

We have seen in previous examples of changing an original string with some operation. We were just generating a new string with every operation because strings are immutable in Python. We can't change a string after they are created. For example, we can't change a string by assigning it to one particular position, but we can create a new string and assign it to the same variable because Python always cleans up an old object.

```
>>> Var = 'Python'
>>> len(Var)
6
>>> Var[-1]
'n'
>>> Var[-2]
'o'
>>> Var
'Python'
>>> Var[1:3]
'yt'
>>> Var[0] = 'z'
Traceback (most recent call last):
  File "<stdin>", line 1, in <module>
TypeError: 'str' object does not support item assignment
>>> Var = 'z' + Var[1:]
>>> Var
'zython'
>>>
```

But, you have one method by which you can change a specific word in the string, but that method is text-based. You can change text-based data if you expand it into individual characters and join it back together or use newer bytearray type available in Python's newer version.

```
>>> Var = 'python'
>>> Var1 = list(Var)
>>> Var1
['p', 'y', 't', 'h', 'o', 'n']
>>> Var1[1] = 'z'
>>> ''.join(Var1)
'pzthon'
```

Every string operation which we have used untill now is like a sequence operation and it can also be used in other Python sequences such as lists and tuples.

String find operation is basic method to search particular substring in string, and string replace method performs replacement of substring in a string. For example,

```
>>> Var = 'python'
>>> Var.find('yt')
1
>>> Var
'python'
>>> Var.replace('py','ze')
'zethon'
```

Here, despite the name of datatype string, we are creating a new string. We are not changing an old string because strings are immutable.

So far, we were understanding the specific operations on the string. But the Python language provides a variety of methods to perform on a string. Some special characters are represented with a backslash. For example, \n is used for end of the line and \t is used for the tab.

You can represent multiple string literals enclosed in triple quotes. Triple quotes are used when you want to concatenate more than one string. For example,

>>>Var = """"python""""programming"""""

Python language comes with full Unicode support, which is required for processing text in international characters like Japanese, Chinese, or other characters which are outside of the ASCII set. You can see non ASCII character sets in web pages, emails, GUIs, or elsewhere. Python has built-in support for Unicode character, but the form varies per Python line.

One point is worth remembering is that Python support pattern-based text processing. Text pattern matching is an advanced tool for Python for beginners, but the readers who have knowledge of other scripting languages knows the importance of pattern matching. This module is used for searching, splitting, and replacement. For example,

Code:

```
>>> import re
>>> match = re.match('Hello[ \t]*(.*)world', 'Hello    Python world')
>>> match.group(1)
'Python '
>>> 
```

The above example searches for the word "Hello" followed by zero or more tabs or spaces, then any character is saved as a match group ending "world". If you find such a substring that matches partial patterns enclosed in parentheses, they are the available group.

Till now, we have studied about numbers and strings in data-types. We are going to study about List, tuple, and dictionary in the upcoming chapter in detail.

Exercise

1. What is a Datatype?

   Answer: The type of data in programming that specifies, what type of value a variable can store such as integer, boolean, string, etcetera.

2. Name fundamental data types present in the Python language.

   Answer:

   - Numbers
   - Boolean
   - String
   - Tuples
   - Lists
   - Dictionary

Chapter 8

Lists and Tuples

Till now, we have learned about different data types and discussed in-details of numbers and strings, which are only two data types in the Python language. Now, we need to understand some more such as Lists and Tuples in detail.

It is really comfortable to deal with the structured format data as the data is set in a specific manner. Python provides data types named "lists" and "Tuples", which is used to organize data in a structured manner. "Lists" and "Tuples" are most popular built-in sequence of the Python language.

Lists:

The Lists are a more flexible ordered collection data type in Python. Unlike strings, lists can contain all type of data such as numbers, strings, and even other lists, too. Lists are mutable in nature so you can change it while assigning and slices.

Properties of Lists:

- Collection of arbitrary objects:
  Lists are the entity where you can collect other objects and treat them as an ordered group. Lists maintain items in left to right positional ordering.

- Accessed by offset:

  In order to access a component, you can fetch any component by indexing the lists. You can fetch it even when it is out of the list. The indexing on object's offset is required for fetching. You can do slicing and concatenation on items because items are set by their position.

- Variable length, nesting:
  Unlike string, lists can grow and shrink according to the need of the program. In addition to that, lists can contain all kinds of objects such as numbers, strings, and another list.

- Mutable:

  You can change lists at any place and it responds to all operations, which are performed on lists like slicing, indexing, and concatenation. It will give result in new lists instead of the new string even if you are changing in a string.

- Object reference:

  Python lists contain zero or more than zero references to the other objects. Whenever you use reference, Python always prefers a reference to an object. For example, you are assigning an object to the data structure component and variable name, then Python will store a reference to the same object name. It will not store the reference to the copy of that object.

Create Lists:

When you want to build a list, you just need to write the number of expressions in square bracket.

Syntax:

> lst_1 = [ ]

lst_2 = [expression1, expression2, ...,

expression N]

For example:

```
list1 = ['script', 'python', 'perl'];
list2 = [1983, 2011];
list3 = [2,4,6, "s", "v", "d"];
```

Access value in Lists:

Lists len(L) always returns the number of items which is present in the list and L[i] represents the items which is at index i and L[i:j] returns a new list which contains objects between "I' and "j".

Code:

```
list1 = ['script', 'python', 'perl'];
list2 = [1983, 2011];
list3 = [2,4,6, "s", "v", "d"];

print ("list1[0]", list1[0])
print ("list3[2:4]", list3[2:4])
```

Output:

```
('list1[0]', 'script')
('list3[2:4]', [6, 's'])
```

Update Lists:

You can add and update single and multiple elements in a list at a time.

Code:

```
list1 = ['script', 'python', 'perl'];

print ("Third value in list is: ")
print (list1[2])

list1[2] = 'programming language'

print ("Updated value in the list is: ")
print(list1[2])
```

Output:

```
Third value in list is:
perl
Updated value in the list is:
programming language
```

Delete elements from Lists:

"del" statement is used for deleting an element from the list.

Syntax:

Del list_name[index_val];

Lists support many operations similar to string. Lists also respond to arithmetic operations same as string, but it will give the result as a new list. For example, + operator will accept the same sort of sequence on both sides. If it is not the same sequence, then it will give type error while compilation.

Code:

```
list1 = ['script', 'python', 1983, 2011];

print list1;
del list1[2];
print "After deleting value at index 2 : "
print list1;
```

Output:

```
['script', 'python', 1983, 2011]
After deleting value at index 2 :
['script', 'python', 2011]
```

If you want to concatenate string and lists, then you need to convert the lists to string to vice-versa.

Code:

```
>>> str([1,2]) + "83"
'[1, 2]83'
>>> [1,2] + list("83")
[1, 2, '8', '3']
>>>
```

If you want to check all sequence operation you have performed in the string, you will see that lists are responding to all sequence operation.

Code:

```
>>> l = ['script', 'python', 'perl']
>>> l[2]
'perl'
>>> l[-2]
'python'
>>> l[1:]
['python', 'perl']
...
```

Indexing and Slicing:

In lists, indexing and slicing work the same as the string because the list is also a sequence. The result of indexing depends on the type of object, which is specified by the programmer at the offset, while slicing always give a new list.

Code:

```
>>> l = ['script', 'python', 'perl']
>>> l[1] = 'java'
>>> l
['script', 'java', 'perl']
>>> l[0:2] = ['program', 'language']
>>> l
['program', 'language', 'perl']
```

Change place in the Lists:

Lists always support the operation which changes the place of the object. Python deals with the object references. The creation of new object and change in place always matters while dealing with a reference because it can impact more than one reference.

While using list, you can change its content by assigning it to the offset or slice.

Code:

```
>>> l = ['script', 'python', 'perl']
>>> l.append('java')
>>> l
['script', 'python', 'perl', 'java']
>>> l.sort()
>>> l
['java', 'perl', 'python', 'script']
```

Both index and slice assignment modify the subject list while dealing with in-place. It will not generate a new lists object. Python list support type-specific method calls. Methods are the function, which is associated with and act upon particular objects. It provides type-specific tools which are generally available for lists.

Tuple:

In the Python language, a tuple is a data type which constructs simple group of objects. You cannot change tuples in place and they are written as a series of items in parentheses, not square brackets.

Properties:

- **An ordered collection of arbitrary object:**

  Tuples maintain left to right order when storing any content. It is a collection of objects which are in a positional order. Tuples can embed all kinds of objects.

- **Access by Offset:**

  You can access items by offset and it supports all operations, which are offset-based such as indexing and slicing.

- **Immutable:**

  Like string, tuples are also immutable. It supports many of the same operations like string and lists. It will not support any in-place change operation, which is applied to the lists.

- **Fixed-length and Nestable:**

You cannot change the size of a tuple without masking a copy because of its immutable property. Tuples can hold any type of object including lists, dictionary, etc. It also supports arbitrary nesting.

- **Object references:**

    Tuple storage access point to other objects and the index tuples are relatively fast.

Create Tuple:

You can create a tuple by comma separated values.

For example,

Tup1 = ('python', 'programming');

Tup2 = (1, 2, 3, 4, 5);

Access Values in Tuples:

You can access the value by using square brackets for slicing along with an index to obtain the value.

Code:

```
tup1 = ('script', 'python', 'perl');
tup2 = (1983, 2011);
tup3 = (2,4,6,8,10,12,14,16);
print ("tup1[0]", tup1[0])
print ("tup3[2:4]", tup3[2:4])
```

Output:

```
('tup1[0]', 'script')
('tup3[2:4]', (6, 8))
```

Updating Tuples:

Tuples are immutable in nature, so you cannot change or update the value of tuples. But you can create a new tuple from an existing tuple and make changes in the new tuple.

Code:

```
tup1 = (1, 2, 3);
tup2 = ('abc', 'def');
tup3 = tup1 + tup2
print (tup3)
```

Output:

```
(1, 2, 3, 'abc', 'def')
```

Delete Tuple:

You can delete tuple by using "del" statement.

Code:

```
tup1 = (1, 2, 3);
tup2 = ('abc', 'def');
tup3 = tup1 + tup2

del tup3;
print (tup3)
```

Output:

```
Traceback (most recent call last):
  File "edit_tuple.py", line 6, in <module>
    print (tup3)
NameError: name 'tup3' is not defined
```

Basic Tuple Operation:

You can use an arithmetic operation like + and * in the tuple. It also supports concatenation and repetition similar to the string and it will give result in a new tuple.

Expression	Result	Description
Len((0, 1, 2, 3, 4))	5	Length
(1, 2, 3) + (4, 5 ,6)	(1, 2, 3, 4, 5, 6)	Concatenation
('Python',)*2	('Python', ' Python)	Repetition
4 in (0, 1, 2)	False	Membership

Indexing and Slicing:

You can operate indexing and slicing similar to string because of its ordered set of the element.

Var = ('python', 'python', 'python language')

Expression	Result
Var[3]	'python language'
Var[-3]	'python'
Var[1:]	['python', 'python language']

If you want to compare elements of two tuples, then you can use 'cmp'.

Syntax:

Cmp(tuple_1, tuple_2)

Description:

tuple_1 = first tuple to be compared.

tuple_2 = second tuple to be compared.

112

If you are comparing elements of the same type, it will give you a direct result, but if you are comparing different types of elements, then you need to cross-check whether it is a number or not. If it is a number, then first perform numeric coercion and then compare them. If they are a string, then it will sorted alphabetically.

Code:

```
tup1 ,tup2 = (123, 'abc'), (456, 'xyz')
print cmp(tup1, tup2);
print cmp(tup2, tup1);
tup3 = tup2 + (789,);
print cmp(tup2, tup3)
```

Output:

```
-1
1
-1
```

If you want to find the length of tuple, then you can use "len()". It will return the number of element in the tuple.

Syntax:

len(tuple)

Description:

Tuple = tuple in which you need to count numbers of elements.

Code:

```
tup1, tup2 = (123, 'abc', 'pqrs'),(456, 'xyz')
print "first tuple length : ", len(tup1);
print "Second tuple length: ", len(tup2);
```

Output:

```
first tuple length :   3
Second tuple length:   2
```

Exercise

1. Explain Lists using Python program.

Answer: The lists are the most flexible ordered collection data type in Python. Unlike strings, lists contain all type of data such as numbers, strings, and even other lists, too. Lists are mutable in nature, so you can change it while assigning and slices.

Code:

```
list1 = ['script', 'python', 'perl'];
list2 = [1983, 2011];
list3 = [2,4,6, "s", "v", "d"]];
```

2. Explain Tuples using Python program.

Answer: In the Python language, a tuple is a data type which constructs simple group of objects. You cannot change tuples in place and they are written as a series of items in parentheses, not square brackets.

Code:

```
tup1 = ('script', 'python', 'perl');
tup2 = (1983, 2011);
tup3 = (2,4,6,8,10,12,14,16);
print ("tup1[0]", tup1[0])
print ("tup3[2:4]", tup3[2:4])
```

Chapter 9

Dictionaries

After string, list, tuple, and numbers, dictionaries is a popular used data type in the Python programming language. It is the last data type to understand in this material. Dictionaries are completely different from other data types. They are not in sequence at all, but still, it is known as mapping.

Mapping is also considered as a collection of other objects, but it stores them as keys instead of their position as the tuple. Mapping doesn't follow any left to right order like tuple; it directly maps keys to associated values.

Properties:

- **Access by keys:** Dictionaries associate a key, so you can fetch an item using the keys from the dictionary. Indexing operation is the same as the list to get component in a dictionary, but the difference between them is it takes the form of the keys and does not use a relative offset.

- **Unordered collection of object:** Items stored in dictionaries are not in order, unlike a list. Keys provide a location of items in a dictionary, but it provides the only symbolic location. It does not provide a physical location, too.

- **Variable length and nesting:** Dictionaries can contain any type of objects and it supports nesting to any depth, too. There can be only one key per key value, but if necessary, the value can be a collection of multiple objects, and a given value can be stored under any number of keys. Dictionaries can grow and shrink without new copies.

- **Mutable:** Dictionaries can be modified by assigning value to indexes, but it does not support sequence operation unlike string and lists because dictionaries are an unordered collection.

- **Object references:** Dictionaries are an unordered table of object references that support access by keys. It is implemented similarly to a hash table, which starts small and grows as per the need. Python uses a optimization hash table algorithm to find the keys — it helps to retrieve data quickly. Like lists, dictionaries also store object references.

Dictionaries Usage:

You can use an arbitrary object such as the standard object or user-defined object in dictionary values. Its values don't have any restrictions on using Python objects, but you cannot use all Python objects with the keys.

There are some points you need to remember about dictionary keys:

- You cannot do more than one entry per key. It means you cannot use a duplicate key. If duplicate keys are encountered during assignment, it takes the last assignment into consideration

- Key should be immutable, which means you can use string, tuples, etc. as dictionary keys, but you cannot use 'key'.

Access Value in Dictionary:

If you want to access elements in the dictionary, you can use the square bracket with the key.

Code:

```
dict = {'Name' : 'Smith', 'Age' : 25}
print (dict['Name'])
print (dict['Age'])
```

Output:

```
Smith
25
```

If you are trying to access elements which are not present in the dictionary, then it will show you an error.

Code:

```
dict = {'Name' : 'Smith', 'Age' : 25, 'Class' : 'Seven'};
print "dict['Mark']: ", dict['Mark'];
```

Output:

```
dict['Mark']:
Traceback (most recent call last):
  File "access_dict.py", line 2, in <module>
    print "dict['Mark']: ", dict['Mark'];
KeyError: 'Mark'
```

Update Dictionary:

You can update a dictionary by adding a new entry, or you can modify or delete an existing entry.

Code:

```
dict = {'Name': 'Smith', 'Age': 10, 'Class': 'Seven'};
dict['Age'] = 14; # update existing entry
dict['School'] = "DPS School"; # Add new entry
print "dict['Age']: ", dict['Age'];
print "dict['School']: ", dict['School'];
```

Output:

```
dict['Age']:  14
dict['School']:  DPS School
```

Delete Dictionary Element:

Dictionary gives you an option of deleting individual elements in the dictionary or deleting entire content, which is present in the dictionary, or you can delete the entire dictionary in a single operation.

You can use "del" statement to remove the entire dictionary.

Code:

120

```
dict = {'Name': 'Smith', 'Age': 10, 'Class': 'Seven'};
del dict['Name']; # remove entry with key 'Name'
dict.clear(); # remove all entries in dict
del dict ; # delete entire dictionary
print "dict['Age']: ", dict['Age'];
print "dict['School']: ", dict['School'];
```

Output:

```
dict['Age']:
Traceback (most recent call last):
  File "delete_dict.py", line 5, in <module>
    print "dict['Age']: ", dict['Age'];
TypeError: 'type' object has no attribute '  getitem  '
```

Dictionary Functions:

1. cmp(dictionary1, dictionary2)

2. len(dictionary)

3. str(dictionary)

4. type(variable)

cmp(dictionary1,dictionary2):

This method is used to compare elements of both dictionaries. It compares both dictionaries based on key and values.

Syntax:

cmp(dictionary1, dictionary2)

Parameters:

121

Dictionary1 = First dictionary to be compared with dictionary2.

Dictionary2 = Second dictionary to be compared with dictionary1.

Code:

```
dict1 = {'Name': 'Zara', 'Age': 7};
dict2 = {'Name': 'Smith', 'Age': 27};
dict3 = {'Name': 'Mark', 'Age': 25};
dict4 = {'Name': 'Adam', 'Age': 10};
print "Return Value : %d" % cmp (dict1, dict2)
print "Return Value : %d" % cmp (dict2, dict3)
print "Return Value : %d" % cmp (dict1, dict4)
```

Output:

```
Return Value : -1
Return Value : 1
Return Value : -1
```

It will return 0 if both dictionaries are equal (dict1 = dict2) in comparison.

If dictionary1 is greater than dictionary2 (dict1>dict2) then it will return 1.

If dictionary2 is greater than dictionary1 (dict2>dict1) then it will return -1.

len(dictionary):

This method gives the length of the dictionary. It counts the number of items and gives the result as a length of the dictionary.

Syntax:

len(dictionary)

Parameters:

Dictionary: Dictionary's length you need to calculate.

Code:

```
dict = {'Name': 'Smith', 'Age': 10};
print "Length : %d" % len (dict)
```

Output:

```
Length : 2
```

str(dictionary):

This method is used to produce a printable string which can represent the dictionary.

Syntax:

str(dictionary)

Parameters:

Dictionary: It is a dictionary.

It will return a string representation.

Code:

```
dict = {'Name': 'Smith', 'Age': 10};
print "Equivalent String : %s" % str (dict)
```

Output:

```
Equivalent String : {'Age': 10, 'Name': 'Smith'}
```

type(dictionary):

This method is used to return the type of variable that you are passing. If passing variable is dictionary then its return type is of dictionary data type.

Syntax:

type(dictionary)

Parameters:

Dictionary: It is a dictionary.

It returns the type of variable that you are passing to the dictionary.

Code:

```python
dict = {'Name': 'Smith', 'Age': 10};
print "Variable Type : %s" %type (dict)
```

Output:

```
Variable Type : <type 'dict'>
```

Sorting keys:

Dictionaries are not in sequence, they don't maintain any left to right order, so when you are printing it, it may come with the different order. If you want all of the dictionary items in proper order, than you can use the dictionary key method to get the key list, sort them by sort method, then iterate through the results in Python for loops. The sorted call returns the result and sorts the various object types sorted in the case dictionary key automatically.

Exercise

1. What is Mapping?

Answer: Mapping is considered a collection of other objects, but it stores them with keys instead of their position. Mapping doesn't follow any left to right order like tuple, it directly maps keys to associated values.

2. Why we need Dictionary?

Answer: You can use an arbitrary object such as the standard object or user-defined object in dictionary values. Its values don't have any restrictions on using Python objects. But, you cannot use all Python objects with the keys.

There are some points which you need to remember about dictionary keys:

- You cannot do more than one entry per key. It means you cannot use duplicate key. If duplicate keys have encountered during assignment, then it takes last assignment into consideration

- Key should be immutable, which means you can use string, tuples, etc. as dictionary keys, but you cannot use 'key'.

Chapter 10

Control Statements

Introduction:

The Python language execution is sequential in nature, but in some cases, you need to change your program's execution sequence based on the problem requirement. Sometimes, you even need to check some conditions, and based on the condition fulfilment, statements need to be executed. To fulfil this requirement, Python provides features like conditional execution, iterative execution, and jumps in the program. They specify the transfer of control from one line to another.

For the conditional execution of statements, Python provides:

1. If-else
2. Switch-case

   For the iterative execution of the code, Python provides:

1. While loop

2. For loop

3. Nested loop

   For jumps in the program, Python has rich features of break and continue. Let's discuss all there feature in detail:

If-else :

This is the most common and powerful feature to implement condition execution of the statement in Python. It is bidirectional in nature. The syntax is as follows:

If expression:

   Statement1

Else:

   Statement2

In the above syntax, Python interpreter evaluates the expression, also called if condition. If the expression results in true (non zero) then statement1 executes. Otherwise, statement2 execution takes place. The following flowchart is suitable for the better understanding of its bidirectional nature:

Flowchart

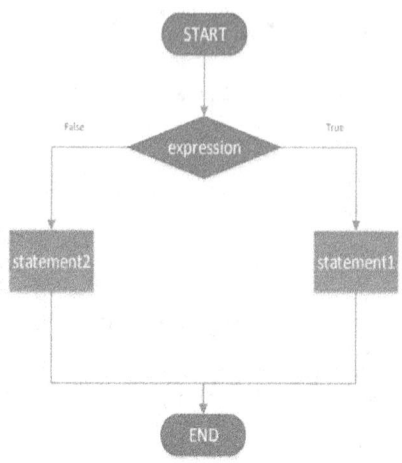

Code

```
"""
Program to check Even and Odd number
"""
num = int(raw_input("Enter a Number :"))
if (num%2==0):
        print "Number is Even"
else:
        print "Number is Odd"
```

Output

```
Enter a Number :5
Number is Odd
```

It is also possible to keep multiple statements with if-else to executes. It just requires putting same indentation space. Block of statements with the same indentation is also called compound statement. The syntax with compound statements is as follows:

If expression:

      Statement1

      Statement2

Else:

      Statement3

      Statement4

FlowChart

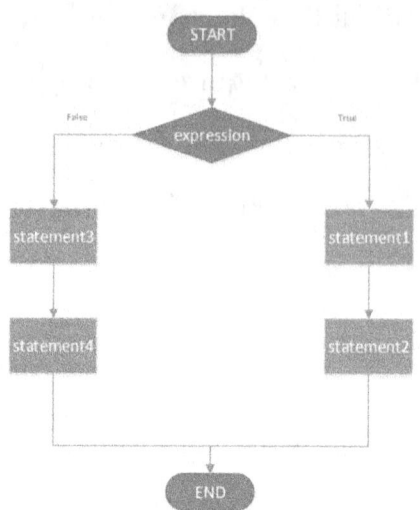

Code

```
"""
Program to test Even and Odd number and display it
"""
num = int(raw_input("Enter a Number :"))
if (num%2 == 0):
        print "Entered number is " + str(num)
        print "Number is Even"
else:
        print "Entered number is " + str(num)
        print "Number is Odd"
```

Output

```
Enter a Number :45
Entered number is 45
Number is Odd
```

Else part of the syntax is not compulsory. You can skip it, according to the need. The syntax and flowchart are as following:

If expression

  statement1

Flowchart3

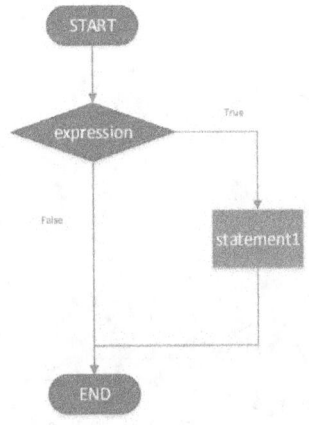

Code

```
"""
Program to test even number
"""
num = int(raw_input("Enter a Number :"))
if (num%2 == 0):
        print "Entered number is " + str(num)
        print "Number is Even"
```

Output

```
Enter a Number :78
Entered number is 78
Number is Even
```

Nested if-else:

The Python language also allows the nesting of if-else where one if-else statement can be used inside the body of other if-else as following:

If expression1:

    If expression2:

        Statement1

    Else

        Statement2

Else

    If expression3:

        Statement3

    Else:

        Statement4

Flowchart

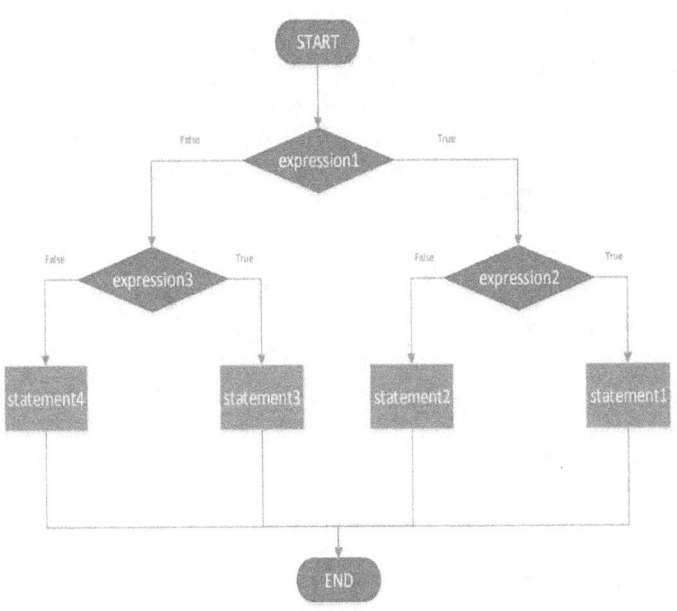

Code

```
"""
Program to find the input range of number
"""
num = int(raw_input("Enter a Number(between 0 to 200):"))
if (num < 100):
        if (num > 50):
                print "Entered number is between 50 and 100"
        else:
                print "Entered number is between 0 and 50"
else:
        if (num > 150):
                print "Entered number is between 150 and 200"
        else:
                print "Entered number is between 100 and 150"
```

Output

```
Enter a Number(between 0 to 200):100
Entered number is between 100 and 150
```

Else-if Ladder:

135

The else-if ladder is one type of multi-way decision-making statement in Python. There is an if-else statement for every else part of if statement and the syntax for the same is as follows:

If expresion1:

    Statement1

Elif expression2:

    Statement2

Elif expression3:

    Statement3

Else:

    Statement4

In the else-if ladder, Python interpreter evaluates every if condition sequentially one-by-one, and when it resolves into true, it executes the corresponding statement and then controls comes out without checking remaining condition.

Flowchart

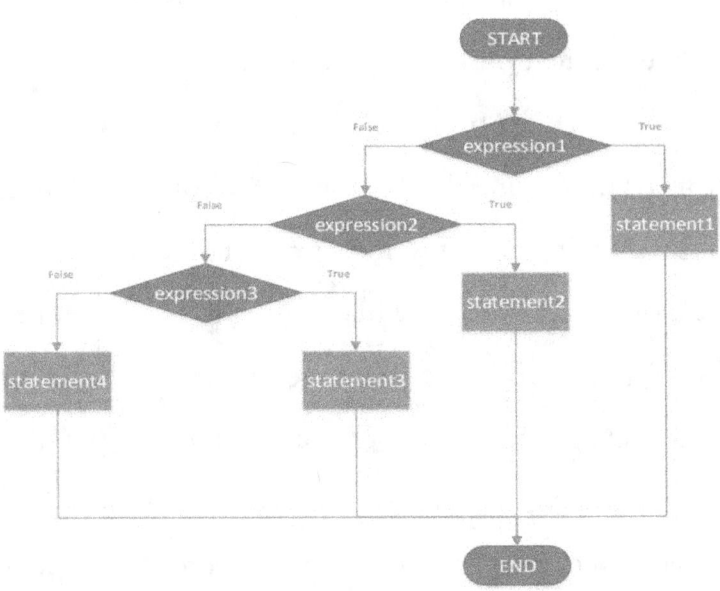

Code

```
"""
Program to test number range
"""
num = int(raw_input("Enter a Number(between 0 to 100):"))
if (num < 25):
        print "Number is between 0 to 25"
elif (num < 50):
        print "Number is between 25 to 50"
elif (num < 75):
        print "Number is between 50 to 75"
else:
        print "Number is between 75 to 100"
```

Output

```
Enter a Number(between 0 to 100):50
Number is between 50 to 75
```

Loops:

137

In any programming language, loops are used when we want to execute a part of program multiple times. It is always easy to optimize the program using Loops. For example, if you want to print a string "Welcome to Python" ten times on the output string, instead of writing print statement ten times, you can use one of the loops (while or for) to implement it. Every Loop in the Python language requires a counter variable, condition check, and increment or decrement operation.

Counter variable keeps track of the number of times the loop has executed. Increment and decrement operation is implemented on the counter variable, and condition check is required for termination of the loop.

Each loop has its own requirement and significance during programming. Let's understand them in detail:

While Loop:

The syntax of while loop is as following:

While expression:
    Statement1
    Statement2

In the above syntax, the expression is evaluated by the interpreter first, and if it resolves into true, then the body of the while loop (Compound Statements) executes. Otherwise, it comes out of the loop. After the execution of the body again, it evaluates the expression and executes the body. The body of loop will execute until the expression in the results into false. This process can be better understood from the below flow chart.

Flowchart

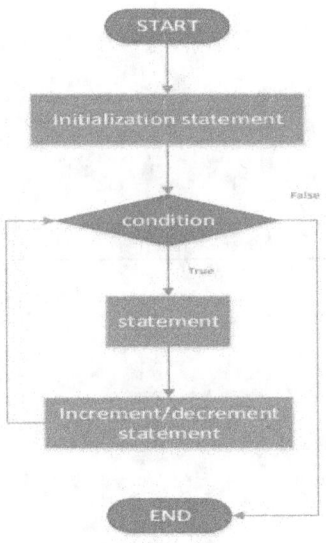

In the programming, you can use the following type of convention for more productive code.

Initialization statement

While condition:

    Statement

    Increment/decrement statement

You will get a clear idea for above convention with the following programming challenge.

Code

```
"""
Program to print 1 to 10 using while-loop
"""
i = 1
while (i <= 10):
        print i
        i = i + 1
```

Output:

```
1
2
3
4
5
6
7
8
9
10
```

For Loop:

The for loop is frequently used out of all the loops because of its easy syntax, which is as follows:

For counterVar in sequence:

    Statements

The syntax comprises of counterVar variable and a sequence. The sequence could be either list, tuple, string, or any collection of data. If you are dealing with data sequencing in Python, for loop is definitely a feasible choice.

During the execution of for loop, the first element in the sequence is assigned to the counterVar and statements of the loop body are executed, then the next element is assigned to the couterVar and the statements are executed in a loop until all the elements of the sequence are exhausted. The sequence in the loop could be any list, string, or collection of data elements.

Flowchart

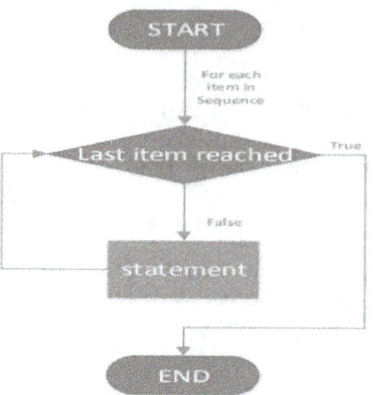

When you need to iterate through the sequence, there are two ways you can iterate using for loop. Let's understand them in brief:

1. Iterating using Sequence expression:
    In this type of for loop, the programmer uses the following syntax to iterate in for loop.

    For element in Sequence:
          Statements

Code

```
"""
Program to print list elements using for-loop
"""
list = ["python","programming","is","fun"]

for list_element in list:
        print list_element
```

Output

```
python
programming
is
fun
```

2. Iterating using Sequence Index:

In this type of for loop, the programmer uses the following syntax to iterate in for loop.

For index in range(len(Sequence)):

> Statements

Code

```
"""
Program to print list elements using for-loop
"""
list = ["python","programming","is","fun"]

for index in range(len(list)):
        print list[index]
```

Output

```
python
programming
is
fun
```

Nesting of loops is also possible by using one loop inside the body of another loop. Application of nested loops can be in the array of sequence and for handling huge data.

Infinite Loop:

The loops that execute its body infinite times are known as the infinite loop. You can implement this type of loop deliberately or by mistake, which puts your program running into continuously. To implement infinite loop, you can use the following approach:

While True:

Statement

The termination of an infinite loop can be controlled by using break and goto statement inside the body of the loop. These statements are explained in the further topics of this chapter.

Continue and Break Statement:

Continue and break statements are very useful statements and used frequently with loops. The syntax for continue is simple:

Continue

Continue statement is used for skipping execution of the loop statements inside the loop body and transferring control to the beginning of the next loop iteration. It is used with the if condition generally. Let's understand its use case with a programming challenge

Code

```
"""
Program to skip numbers using 6 to 10 using continue statement
"""
i = 0
while i<10:
    i = i + 1
    if i == 5:
        continue
    print i
```

Output

```
1
2
3
4
6
7
8
9
10
```

The break statement is similar to the continue, but when it is used inside the loop, it terminates the loop and control is transferred to the next statement after the loop. Let's understand it with following program.

Code

```
"""
Program to stop while-loop using break statement
"""
i = 0
while (i<10):
        i = i + 1
        if (i > 5):
                break
        print i
```

Output

```
1
2
3
4
5
```

Exercise

1. What is the importance of loops in programming?

   Answer: Defining a loop in your code allows the computer to repeatedly perform certain tasks. Depending on the task to be performed, the loop needs to be defined in the computer program for a variety of reasons. The computer programming language needs to be looped so that the code executes actions as many times as needed.

2. Name different types of available loops in Python.

   Answer:

   - For loop
   - While loop
   - Infinite loop using for and while loop

Chapter 11

Functions and Modules

Introduction:

Throughout the previous chapters, we have discussed the different features of Python interpreter that will help you to create your Python program. It's time to move to the design approaches for your programs and without an understanding of the functions and modules, it would be impossible to create a properly designed program. Functions and Modules give you the freedom to cut your program into small parts and implement it with an easy-to-design philosophy.

Functions and its Uses:

In simple words, a function is a collective group of Python statements. The ideology behind the use of functions is to reuse the code. Whenever you come across a situation where you want to execute a group of statements more than once, then you need to create a function. It is a programming practice to write a function and call it with its name every time. You can also perceive functions as the independently running programming section, that you can use multiple times.

Functions are like devices that have the capability of taking input parameters and provide output. Output of the function can be either a data or operation on the parameter passed in it.

Before we dig deep into the syntax and programming with functions, let's understand a bigger picture for the use case of functions. Functions are generally giving a structure to your Python program. Sometimes they are also called procedures and sub-routines in other programming languages. Primarily, there are following philosophy for the use of functions in any python program:

1. Maximum Code-reuse and Minimum redundant programming:
   It is similar to any other programming language. Functions are the easiest way to package your Python logic, you just need to write your code logic once in the

function body, and later you can use it multiple times in your program. It also minimizes your redundant statements.

2. Well-structured programming:

   The function gives you a tool to divide your big programming task into multiple well-defined procedures and allow you have a well-structured program for the same. Let's consider a programming scenario where you want to calculate average salary of the employee in any organization. You can divide the task into procedures likes taking the input of the employee data, calculating an average, and displaying the average value. The function can be written for each of the procedures and call them to have the well-structured program.

Function Syntax:

In Python programming, the general syntax of writing function is as follows:

```
def functionName( arg1, arg2 …. argN):
    Statements
    Return val
```

def is considered as the header of the function, which generates a function object and assigns a function name to it. In the brackets, the function multiple input parameters are represented with arg1, arg2 … argN. These arguments are optional when the function does not take any input parameter then brackets are kept empty. After the colon, function bodies with multiple statements is written where functional logic is implemented. The return statement returns val value to the caller in the program. It can appear anywhere inside the function body and usually is present at the end of the function. If val is not specified, then function returns None as the return value. Both the val and return statement are optional.

Let's get into the Python programming to get more use out of case of functions and its implementation:

Code

```
"""
Program to add two numbers using a function
"""
def main():
        num1 = 10
        num2 = 20
        num3 = add( num1, num2)
        print "Addition is :" + str(num3)

def add( a, b):
        c = a + b
        return c

if __name__ == "__main__":
        main()
```

Output

```
Addition is :30
```

Code

```
"""
Program to test even odd using function
"""
def main():
        input_num = int(raw_input("Enter a Number :"))
        evenOdd(input_num)

def evenOdd(num):
        if (num%2 == 0):
                print "Number is Even"
        else:
                print "Number is Odd"
        return

if __name__ == "__main__":
        main()
```

Output

```
Enter a Number :45
Number is Odd
```

Modules and its Uses:

Modules are the top level programs which organize programming units. It contains packages that have Python code, reusable data, and namespaces, which reduces clashing of variables in your Python program. In a simple way, modules can be considered as the program files. And every file which is referred is called as a Module.

Modules are generally processed by using import and from statement. Let's understand about these statements before digging into deep into modules:

153

1. Import:

> It allows you to load complete module as a whole in your Python program.

2. From:

> It allows you to load specific names from any module in your Python program.

As any particular module is being loaded inside your Python program, it lets you use all the self-contained program codes from the modules. Because of the use of modules inside Python programming, it provides you with a bigger picture with the use of existing modules without any conflicts between attributes and methods.

There are many uses of modules, let's understand them in a brief:

1. Code – reuse:

When you are loading any of the modules in your Python using import statement, you can use all the methods and functions present in the particular module. After importing, it can be referenced multiple times to reduce the lines of code. Modules always help to visualize a bigger picture of the program. Unless you are using Python interpreter, you can import modules just by using its name.

2. Separate Namespaces:

As modules are a self-contained program code, being a programmer, their parameters are isolated from your main Python code. It helps you to write your Python code in a well-organized manner, keeping top-level organization in mind.

Whenever you are working with Python programming, you will need to import and link libraries with your main top-level program. Libraries are present inside the module files, which act as a tool to perform programming tasks.

Let's understand the concept of modules and its use with programming examples. There are following files with their Python code:

```
Def display(text):          #displayModule.py

    Print text
```

```
Import displayModule                        #
mainScript.py

displayModule.display("Hello, World!!")      # prints
"Hello, World!!"
```

In the above example, "mainScript.py" is a top-level file that contains text in it. The execution of the top-level file occurs in a top to bottom. And "displayModule.py" are modules files containing def statements and assigns function object to the name "display". Inside the function body, print statement is present and displays the passing parameter to the output screen.

The top-level files include an import statement that loads the modules into the main file. After fetching modules, it can be referenced using the attributes of it.

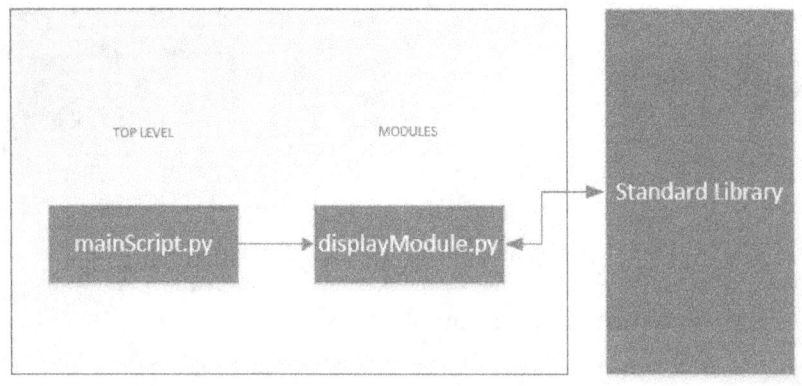

Exercise

1. What is Function and write uses of function in a programming language?

   Answer: In simple words, a function is a collective group of Python statements. The ideology behind the use of functions is to reuse the code. Whenever you come across a situation where you want to execute a group of statements more than once, then you need to create a function. It is a programming practice to write a function and call it every time with its name. You can also perceive functions as the independently running programming section, which you can use multiple times.

   Functions are like devices that have the capability of taking input parameters and provide output. The output of the function can be either a data or operation on the parameter passed in it.

2. What are modules and its uses?

Answer: Modules are the top level programs that organize programming units. It contains packages that have Python code, reusable data, and namespaces which reduces clashing of variables in your Python program. In a simple way, modules can be considered as the program files, and every file, which is referred is called as a Module.

- When you are loading any of the modules in your Python using import statement. You can use all the methods and functions present in the particular module. After importing, it can be referenced multiple times that reduce the lines of code. Modules always help to visualize a bigger picture of the program. Unless you are using Python interpreter, you can import modules just by using its name.

- As modules are self-contained program codes, being a programmer, their parameters are isolated from your main Python code. It helps you to write your Python code in a well-organized manner with top-level organization in mind.

Chapter 12

File Input-Output

Each program is a combination of program statements to perform some task or logic. These logics may or may not require inputs to provide the output, hence inputs that are outputted are part of every program. You need a file to store everything for storage on the computer, which is managed by OS. Although variable provides us a way to store the data while a program is running, we must save it to a file if we want to keep the data after the program has ended.

There are always two parts of a computer system's file, one is a filename and another is an extension of the file. In addition, these files also have two key attributes, which are name and location or path which specify the location of file in the computer. The two parts of the filename are separated by dots (.) or periods.

The built-in open method is used to create a Python file object that provides a connection to the files, which resides on the programmer's machine. After calling an open function, the programmer can transfer the data string to and from an external file residing in the machine.

Print to the screen

You can produce output by using the "print" statement where you can pass expression separated by commas. This function converts the expression which you are passing into a string and writes the result to standard output.

<u>Code:</u>

```
print "python is widely used programming language";
```

<u>Output:</u>

```
python is widely used programming language
```

Read Input:

You can read a line of text from standard input, which will come from the keyboard by using two built-in functions.

- raw_input

- input

- **raw_input function:**

  The raw_input function reads one line from standard input and returns output as string.

Code:

```
str = raw_input("Enter string: ");
print "input is: ", str
```

Output:

```
Enter string: Hello python
input is:  Hello python
```

- **input function:**

  The input function assumes that the input is a valid Python expression and it will return the evaluated result to you.

Code:

```
str = input("Enter string: ");
print "input is: ", str
```

Output:

```
Enter string: [x*5 for x in range(2,10,2)]
input is:  [10, 20, 30, 40]
```

Open and close the file:

From the beginning of this chapter, we understood the function related to input and output from the users. In the continuing discussion, we will have an in-depth understanding of taking inputs from the file and storing the output to it.

Open Function:

You need to open file before you start reading and writing any file. Python has a built-in function that is used to open file i.e. open(). This function will create a file object, which is utilized to call other methods associated with it.

Syntax:

File object = open(file_name[, access_mode][, buffering])

Parameters:

file_name = The file_name is a string value which contains a file name to access.

access_mode = The access mode provides the mode in which the programmer wants to open the file i.e. read, write, append etc.

buffering = If buffer value is 0, that means no buffering. If it is 1, then line buffering is performed while accessing the file. If you specify the buffer value as a greater than 1, then buffer the operation execute with the specified buffer size. If it is negative, the buffer size is system default.

Different Modes:

Modes	Description
R	Open file for read only
r+	Open file for both read and write
rb	Open file for read in binary format
rb+	Open file for both read and write in a binary format
W	Open file for write only
w+	Open file for both read and write
wb	Open file for write in a binary format

wb+	Open file for both read and write in a binary format
A	Open file for appending
a+	Open file for both appending and reading
ab	Open file for appending in binary format
ab+	Open file for both appending and reading in binary format

Close Function:

The close () method of the file object refreshes any unwritten information and closes the object file and the object cannot be written later.

Python closes the file automatically when the file is reassigned to another file.

Syntax:

fileObject.close();

Code:

```
foo = open("python.txt", "wb")
print "Name of the file: ", foo.name
foo.close()
```

Output:

```
Name of the file:  python.txt
```

Read and Write the file:

Write Function:

You can write any string to an open file by using write () function. It is really important that Python string contains binary data and not just text. It does not add a new line character to the end of the string.

Syntax:

fileObject.write(string);

Code:

```
foo = open("python.txt", "wb")
foo.write( "Python is a widely used programming language.\nYeah its great!!\n");
foo.close()
```

Output:

```
Python is a widely used programming language.
Yeah its great!!
```

The above method will create .text file and writes content in the file, and after execution, it closes the file.

Read Function:

You can read a string from an open file by using read ()
function.

Syntax:

fileObject.read([count]);

Parameters:

The passing parameter is representing the number of bytes to be read from the open file. This method starts reading from the beginning of the file, and if count is missing, then it tries to read as much as possible, maybe until the file is over.

Code:

```
foo = open("python.txt", "r+")
str = foo.read(10);
print "Read String is : ", str
foo.close()
```

Output:

```
Name of the file:  python.txt
Closed or not :  False
Opening mode :  wb
Softspace flag :  0
happy@happy-300E4C-300E5C-300E7C:~/python$ gedit open.py
happy@happy-300E4C-300E5C-300E7C:~/python$ gedit close.py
happy@happy-300E4C-300E5C-300E7C:~/python$ python close.py
Name of the file:  python.txt
happy@happy-300E4C-300E5C-300E7C:~/python$ gedit close.py
happy@happy-300E4C-300E5C-300E7C:~/python$ gedit write.py
happy@happy-300E4C-300E5C-300E7C:~/python$ python write.py
happy@happy-300E4C-300E5C-300E7C:~/python$ open write.txt
Couldn't get a file descriptor referring to the console
happy@happy-300E4C-300E5C-300E7C:~/python$ gedit python.txt
happy@happy-300E4C-300E5C-300E7C:~/python$ gedit write.py
happy@happy-300E4C-300E5C-300E7C:~/python$ gedit read.py
happy@happy-300E4C-300E5C-300E7C:~/python$ gedit python.txt
happy@happy-300E4C-300E5C-300E7C:~/python$ python read.py
Read String is :  Python is
```

File Position:

If you want to check current position with the file, then you can use *tell()* function. The next read and write will occur after the number of bytes returned from the tell () function from the beginning of the file.

The *seek* (offset [, from]) is used to change the current file position. The offset indicates the number of bytes to move. The *from* is used to specify the reference position from which you want to move the bytes.

169

If from is set to 0, the beginning of the file is used as a reference position. 1 indicates that the current position is used as a reference position. If it is set to 2, then the end of the file will be treated as a reference position.

Code:

```
foo = open("python.txt", "r+")
str = foo.read(10);
print "Read String is : ", str
position = foo.tell();
print "Current file position : ", position
position = foo.seek(0, 0);
str = foo.read(10);
print "Again read String is : ", str
foo.close()
```

Output:

```
Read String is :  Python is
Current file position :  10
Again read String is :  Python is
```

Rename and Delete File:

Rename Function:

Rename function generally takes two arguments i.e. current filename and new filename.

Syntax:

os.rename(current_filename, new_filename)

Remove Function:

You can delete files by giving the name of the file as an argument in the remove () function.

Syntax:

os.remove(file_name)

Code:

```
import os
# Remove a file python1.txt
os.remove( "python1.txt" )
```

File Flush:

Python automatically flushes the files when it is closed. But if you want to flush the data before closing the file, then you can use flush () function. This method is used to flush the internal buffer.

Syntax:

fileObject.flush();

It does not return any value.

Code:

```
foo = open("python.txt", "wb")
print "Name of the file: ", foo.name
# Here it does nothing, but you can call it with read operation.
foo.flush()
foo.close()
```

Output:

```
Name of the file:  python.txt
```

File next:

The *next* () function is used when the file is used repeatedly or iteratively. It returns the next input line and raises *StopIteration* when end of the line hits.

Using the next () method with other file methods such as *readline* () is not correct. However, using *seek* () to relocate the file to an absolute position refreshes the read-ahead buffer.

Syntax:

fileObject.next();

Next () function will return the next input line.

Code:

```python
foo = open("python.txt", "rw+")
print "Name of the file: ", foo.name
# Assuming file has following 3 lines
# This is 1st line
# This is 2nd line
# This is 3rd line
for index in range(3):
    line = foo.next()
    print "Line No %d - %s" % (index, line)
foo.close()
```

Exercise

1. What is the usage of help () and dir () function in Python?

   Answer: The Help () and dir () functions can be accessed from the Python interpreter and used to view merge dumps of built-in functions.

   - Help Function: The help () function is used to display document strings, as well as help with modules, keywords, properties, and more.
   - Dir Function: The dir() is used to display the symbols which is defined.

2. What are negative indexes and where it is used?

   Answer: The sequence in Python is indexed and consists of positive numbers and negative numbers. The positive numbers use '0' as the first index and '1' as the second index, so the process is done.

The negative index begins with '-1', indicating the last index in the sequence, '-2' as the penultimate index, and the sequence going forwards like a positive number.

Chapter 13

Object-oriented Programming

Introduction:

The secondary philosophy behind the development of the Python language was to create an easy-to-code object-oriented programming language that has the capability of less development time with all the advantages of object-oriented. Though using Python's object-oriented way of programming is optional, but it is a good practice over procedural programming.

You can certainly use procedural programming practice with Python, which allows you to develop pretty quickly. In practice, Object-oriented programming requires a lot of pre-planning in the actual development of the solution, hence it is used for the large projects. When the time for the solution development is less, then top-bottom approach in writing Python scripts are a better option. In some situations, if the pre-planning and program modelling strategies are properly formed for larger projects, then development time could be significantly reduced.

If you are not familiar with basics or object-oriented fundamentals, then it is advisable to refer all the basic principles of object-oriented programming. Before getting deep in the object-oriented programming, let's get familiar with various terminologies associated with it:

1. Class:

   The class is a prototype, which is user-defined and specifies a standard set of attributes. These attributes are methods, instance variables, and data variables.

2. Class Variable:

   Class variables are the object or variables which are shared in a particular class. These variables are declared and defined inside the body of a class, but outside of method present in the class. Generally, these types of variables are less commonly used than instance variables.

3. Instance:

A specific object class is called an Instance of that particular class.

4. Instance Variable:

The variables which are declared and defined inside the body of the class method and its scope are only inside the method body.

5. Object:

An object is the basic building block of any object-oriented programming language. It is a particular instance of the data structure that is defined by its class. The object includes methods instance variables and class variables.

6. Method:

The method is a small function or procedure defined inside a class. These are the building blocks of any class that implements certain logic.

7. Inheritance:

Inheritance is one of the popular advantages of using an object-oriented programming language. It is a process in which the characteristics of a class is transferred to the other class. The new class, which is derived from the former class, is also known as the child class.

Now, let's get started with the object-oriented programming in further sections:

Creating a Class:

Classes are the user-defined prototypes with its attributes. To create a class in the Python language. The following syntax is used:

class ClassName:
      "Class Documentation string"
      classAttributes

In the above syntax, the class is a statement that creates a class with class name as className. The next line after the colon is for documentations of class. The documentation string contains all the information about the class in the double inverted comma. The class body has classAttributes and it comprises of class variables, instance variables, and methods.

Example of Class

Let's understand the fundamental of object oriented class with a simple programming example.

Code

```
"""
Program to create Employee Class
"""
class Employee:
        'Base class for Employee'
        employeeCount = 0;

        def __init__(self, name, salary):
                self.name = name
                self.salary = salary
                Employee.employeeCount += 1

        def displayCount(self):
                print "Total Employees are %d" % Employee.employeeCount

        def displayEmployee(self):
                print "Name : ", self.name, ",Salary :", self.salary

# Creating fist object of Employee Class
employee1 = Employee("Alex", 8000)

# Creating second object of Employee Class
employee2 = Employee("Neo", 10000)

# Displaying employee1 and employee2 data
employee1.displayEmployee()
employee2.displayEmployee()

# Displaying totoal number of employee
print "Totol Employee :%d" % Employee.employeeCount
```

Output

```
Name :   Alex ,Salary : 8000
Name :   Neo ,Salary : 10000
Totol Employee :2
```

In the above example, Employee class can have multiple attributes such as Employee name, Employee salary, and their count, hence class allows the programmer to specify

180

the entity with its features. displayCount and displayEmployee are the methods of the Employee class. Inside the Employee class, employeeCount variable is instance variable as its scope is inside the class only.

The method name with __init__ inside the Employee class is called the constructor or initialization method whenever object of Employee class is created, then its attributes are initialized with the specified arguments.

To create an object of the class, it can be called with its name and initialized parameter is passed. In the above program, employee1 and employee2 are two objects of Employee class. To access the attributes of any class, it can be used with className, dot operator, and attribute name. As you can see, to call displayEmployee, method employee1.displayEmployee() is used.

The object-oriented programming philosophy helps in distributing the real time entities as classes and allows the programmer to write modular code and implement it for larger applications.

Exercise

1. What is object oriented programming?

   Answer: OOPS is abbreviated as an object-oriented programming system, in which programs are treated as a collection of objects. Each object is an instance of a class.

2. Explain function overloading?

   Answer: Function overloading is defined as a normal function, but it has the ability to perform different tasks. Through the function input and output types, you can create several methods with the same name.

Chapter 14

Code Optimization

Python is one of the most popular and widely used programming languages for solving programming challenges. There can be many solutions for the particular problem by using different logics, but the effectiveness of any solution is measured in terms of time and memory consumed. If your solution is giving correct output but taking a long time to run, then it is not optimized, it is similar to memory consumption. Your program should be consuming optimum memory. But there is always a trade-off between these two parameters. Because when you try to write high-speed code, then it increases memory consumption of the system and vice versa, but based on the application requirements, one can find a well-optimized solution.

Creating a highly effective solution takes a lot of programming experience and in-depth knowledge of the

Python language. In the further section of this chapter, we have discussed some techniques for finding an optimized solution, they are as following:

- **Use built-in function and library:** Built-in function is really helpful for optimizing any code. The interpreter does not need to execute particular loops so it will give you fast results.

  The packages are platform specific, which means if you are doing string operation, then it is better to use Python packages to optimize your code. For example, use existing module "collection" like "deque" which is an optimized way while dealing with strings.

  ### Code:

  ```
  from collections import deque
  s = 'python'
  d = deque(s)
  d.append('y')
  d.appendleft('h')
  print d
  d.pop()
  d.popleft()
  print list(reversed(d))
  ```

  ### Output:

  ```
  deque(['h', 'p', 'y', 't', 'h', 'o', 'n', 'y'])
  ['n', 'o', 'h', 't', 'y', 'p']
  ```

- **Sort using keys:** You can use the key parameter of built-in sorting, which is a faster way to sorting

Code:

```
list = [1, -3, 6, 11, 5]
list.sort()
print list

s = 'python'
s = sorted(s)
print s
```

Output:

```
[-3, 1, 5, 6, 11]
['h', 'n', 'o', 'p', 't', 'y']
```

- **Optimize loop:** You should write your code with timing parameters in your mind, particularly when dealing with loops. Because Python is designed to have only one way to do task.

Code:

```python
s = 'pythonprogram'
slist = ''
for i in s:
    slist = slist + i
print slist

# string concatenation
st = 'pythonprogram'
slist = ''.join([i for i in s])
print slist

# Better way to iterate a range
evens = [ i for i in xrange(10) if i%2 == 0]
print evens

# Less faster
i = 0
evens = []
while i < 10:
    if i %2 == 0:
        evens.append(i)
        i += 1
print evens

# slow
v = 'for'
s = 'python ' + v + ' python'
print s

# fast
s = 'python %s python' % v
print s
```

Output:

```
pythonprogram
pythonprogram
[0, 2, 4, 6, 8]
```

- **Try multiple methods in coding:** Always try multiple approaches while creating an application because one may give you better results than another. For the different inputs, it takes different times for execution. For some particular set of inputs, your chosen solution may be slow, you can decide as per your application need.

Code:

```
my_dict = {'p':1,'r':1,'o':1,'g':1}
word = 'pythonprogram'
for w in word:
    if w not in my_dict:
        my_dict[w] = 0
    my_dict[w] += 1
print my_dict

# faster
my_dict = {'p':1,'r':1,'o':1,'g':1}
word = 'pythonprogram'
for w in word:
    try:
        my_dict[w] += 1
    except KeyError:
        my_dict[w] = 1
print my_dict
```

Output:

```
{'a': 1, 'g': 2, 'h': 1, 'm': 1, 'o': 3, 'n': 1, 'p': 3, 'r': 3, 't': 1, 'y': 1}
{'a': 1, 'g': 2, 'h': 1, 'm': 1, 'o': 3, 'n': 1, 'p': 3, 'r': 3, 't': 1, 'y': 1}
```

- **Use *xrange*:** This function is used to display a number by looping because it returns the generator object. This function is used to display only particular range on demand and hence it is known as "lazy evaluation". But it can save your system memory because it will yield only integer element at a time.

Code:

```
# slower
x = [i for i in range(0,10,2)]
print x

# faster
x = [i for i in xrange(0,10,2)]
print x
```

Output:

```
[0, 2, 4, 6, 8]
[0, 2, 4, 6, 8]
```

- **Use local variable:** Python retrieves local variable faster than retrieving global variable. Avoid global variable as much as you can. If you are accessing any statement often, which is inside a loop, then write it to a variable.

Code:

```
# run faster
class Test:
    def func(self,x):
        print x+x

Obj = Test()
my_test = Obj.func # Declaring local variable
n = 2
for i in range(n):
    my_test(i) # faster than Obj.func(i)
```

- **Lambda Function:** Lambda function is an anonymous function that can be used with *filter* (), *map* () and *reduce* () function.

Code:

```
>>> f =lambda x,y: x/y
>>> f(1,1)
1
>>>
```

Filter () –

Syntax:

               filter (function, list)

First parameter of "filter" is function and another is list.

Code:

```
>>> f =lambda x,y: x/y
>>> f(1,1)
1
>>> a = [1,2,3,4]
>>> p = map(lambda x:x*10, a)
>>> print p
[10, 20, 30, 40]
>>> a = [10,20,30,40,50,60]
>>> p = filter(lambda x: x % 2,a)
>>> print p
[]
>>> a = [1,2,3,4,5,6,7,8,9,10]
>>> p = filter(lambda x: x % 2,a)
>>> print p
[1, 3, 5, 7, 9]
>>> ▮
```

Map () –

Syntax:

map (function, list)

First parameter of "map" is function and another is list.

Code:

```
>>> f =lambda x,y: x/y
>>> f(1,1)
1
>>> a = [1,2,3,4]
>>> p = map(lambda x:x*10, a)
>>> print p
[10, 20, 30, 40]
>>> ▮
```

Reduce () –

Syntax:

reduce (function, list)

First parameter of "reduce" is function and another is list.

Code:

```
>>> a = range(2,6)
>>> p = reduce(lambda x,y:x+y, a)
>>> print p
14
>>>
```

- **List:** Use list instead of lengthy code. As it gives you the flexibility to eliminate a large number of lines from the program

Code:

```
>>>    q = [ ]
   File "<stdin>", line 1
     q = [ ]
     ^
IndentationError: unexpected indent
>>> q = [ ]
>>> for i in range(5,10):
...     for j in range(i*2,20):
...         q.append(j)
...
>>> print q
[10, 11, 12, 13, 14, 15, 16, 17, 18, 19, 12, 13, 14, 15, 16, 17, 18, 19, 14, 15,
 16, 17, 18, 19, 16, 17, 18, 19, 18, 19]
>>>
```

Optimized way:

```
>>> a= [j for i in range(5,10) for j in range(i*2,100)]
```

191

- **Dictionary:** Use dictionary comprehension for optimization while creating a dictionary.

  Code:

  ```
  >>> d = {k: k*3 for k in range(1,5)}
  >>> print d
  {1: 3, 2: 6, 3: 9, 4: 12}
  >>>
  ```

- **Use Import in proper manner:** Sometimes you need a particular package for a particular module so it is an optimized way if you specify particular package and module.

  Code:

  ```
  Normal way: from country import *
  Correct way: from country.india import states
  ```

- **Lazy Generator:** If you are using range for finding some of 100 elements, then it will be waste of memory. You can use *xrange* for optimization, as it generates each number in which sum will consume to accumulate the sum.

  Code:

```
>>> n=sum(range(100))
>>> print n
4950
>>>
```

- **Peephole Technique:** It is a technique which is used to optimize small segments of instruction from a program. The segment is called as 'Peephole' or 'window'. It spots the instructions you can replace with minified program or instruction.

Code:

```
>>> ele = 'peephole'
>>> if ele in {'peephole', 'demo', 'optimization'} : print("TRUE")
...
TRUE
>>>
```

In this example, we used the "in" operator to find particular elements from the collection. Here, Python detects that the collection will be used to verify the membership of the element. So it treats these instructions as a constant operation regardless of the size of the collection and it processes faster than tuples and lists. This method is also known as membership test in Python.

- **Use Advance profile with C Profile:** C profile is a part of packages in the Python programming. You can use C profile in many ways with your Python code. For example, you can wrap a function inside run method to measure performance of the program or run the script from command line with c profile as an argument.

Code:

```
>>> ele = 'peephole'
>>> if ele in {'peephole', 'demo', 'optimization'} : print("TRUE")
...
TRUE
>>> import cProfile
>>> cProfile.run('10*10')
        2 function calls in 0.000 seconds

  Ordered by: standard name

  ncalls  tottime  percall  cumtime  percall filename:lineno(function)
       1    0.000    0.000    0.000    0.000 <string>:1(<module>)
       1    0.000    0.000    0.000    0.000 {method 'disable' of '_lsprof.Prof
iler' objects}

>>> █
```

You can look at a result and find out the area where you think you need to improve. You can attach C profile while running script too.

- **Interpret C Profile result:** It is even more important to find the culprit in analyzing the output. If you are able to find key element which constitute the CProfile report, then only you can make decision.

1. ncalls – Number of calls made.

2. tottime – time spent in given function.

3. percall – Represent quotient of "tottime" divided by "ncalls".

4. Cumtime – cumulative time in executing function.

5. filename_lineno (function) – Point of action in a program.

- **Optimization using IF statement:** Most of the programming languages allow for laziness – if evaluated, Python does, too. This means that if you add the "AND" condition, not all of the conditions will be tested when anyone is true unless it is an error.

   You can utilize this technique by normal adjustments of your current code. For example, if you are searching for a specific pattern in a program then you can reduce the scope with the use of "AND" condition.

Exercise

1. How does memory management works?

   Answer: Python memory is managed by Python's private heap space. All Python objects and data structures are in a private heap. Programmers do not have permission to access this private heap; the interpreter is responsible for handling this Python private heap.

   The Python heap space allocation for Python objects is done by the Python memory manager. The core API provides some tools for programmers to write code. Python also has a built-in garbage collector that reclaims all unused memory and frees memory and makes it available for heap space.

2. Why all memory is not de-allocated in Python?

   Answer: Whenever Python exits, especially those Python modules that have circular references to other objects or objects referenced from the global namespace are not always de-allocated or freed. On exit, due to its own efficient cleanup mechanism, Python will try to release / destroy all other objects.

Chapter 15

Useful Python Libraries

Throughout this book, we have discussed various features of the Python language and its utilities, there are almost limitless uses of Python currently. Its uses in the various domains are due to its quick and easy programming approaches. Various libraries present until the date, enrich its usability. You could name any domain for programming and its libraries are available on the internet. It is just matter of importing those libraries in your code and using its modules for your program application.

We already discussed Python libraries related to mathematical and scientific application in "Mathematical Aspects". Let's see some more Python libraries and its applications a more in-depth. Meanwhile, you will get to know many ways to use it for your programming tasks. Let's get started!

Tkinter Library:

This library is built-in present with all the Python packages, so you don't need to install it separately on your system. As we have discussed in the chapter "_____". Tkinter's name is shorthand name for interface to Tk. This is one of the many GUI libraries for Python. To import this library into your program, you can use the following line:

$ import Tkinter

Or

$ from Tkinter import *

In case if you want to include only some module from it, you can use:

$ from Tkinter import moduleName

Where moduleName is any module name present in the Tkinter, the available modules are discussed in the further sections.

Uses:

You can use it to create your Graphical User Interfaces such as forms, button, checkboxes, and many other GUI features. Front-end designing is important aspects when creating any application. This library helps you in it.

Modules:

ScrolledText: to create a text widget having a scrollbar with it.

tkColorChooser: It allows the user to select a particular color.

tkCommonDialog: to create dialog box of different types.

tkFileDialog: To provide a dialog box to select or save the file by the user.

tkFont: To use the different font for GUI.

tkMessageBox: To create message boxes.

tkSimpleDialog: It provide primary dialog box functions.

PyQT Library:

PyQt is the most popular graphical user interface libraries, which is developed by Riverbank Computing Ltd. The library is used not only for computer applications, but also in embedded applications. There are many version of PyQT has been released.

To import this library in your code, you need to install PyQT using pip installer on your system.

Uses:

The uses of PyQT library are diverse. Some of the very complex applications (including Embedded Applications) using it for the development of their graphical user interface.

Modules:

There are hundreds of modules available from PyQT library; you just need to use particular modules as per your applications. You can go to the link for exploring its documentation:

http://pyqt.sourceforge.net/Docs/PyQt5/

Some of the general PyQt modules are as follows:

QtGui: It is used for system integration, handling GUI events, 2-dimensional graphics, basic images, text, and fonts.

QtWidgets: This module comprises of almost all classic user interface elements such as button, textbox, list wheel, etcetera.

QtFileDialog: This module contains all the classes and function related to selection and saving of files by the user.

Requests Library:

Requests is a very simple and quick HTTP library which was developed by Kenneth Reitz. It is the must known library for any Python programmer. Its beloved features attract every web Python developer.

To install it on your system and use it with your Python program, you need to setup it using pipenv.

Uses:

This library is useful for requesting URL in an automated way. There are various features available with it, such as network pooling, connecting to international domains and URLs, browser type SSL verification, and automatic decoding of content.

Modules:

Request: this method helps in sending a request to URL specified with it.

Head: this method is used to send the HEAD request.

Get: this method sends a GET Request.

Put: this method helps in sending PUT request.

Patch: this method is used to send PATCH request to URL.

Delete: this method is used to send DELETE request.

Exception:

There are many exceptions that occur while working with requests library. Let's understand these exceptions and there causes:

RequestException: Whenever there are ambiguous exceptions during the request.

ConnectionError: Whenever there is connection error occurs.

URLRequired: Whenever the correct URL is required for requesting.

ConnectionTimeout: Whenever a timeout occurs when connecting to a remote server.

HTTPError: Whenever HTTP error occurs.

SQLAlchemy Library:

It is one of the important Python database access libraries. It includes all the tools required for accessing SQL database and mapping to it. It provides flexibility and power to the developer for writing high-performance and efficient database program.

There are many advanced level database access functions available in this module. To install this library on your system, you need to take help from pip module.

Uses:

The fundamental utility of SQLAlchemy is to link your Python application to SQL database and access it using all the powers of SQL. The most popular feature of the SQLAlchemy library is ORM (Object-Relational Mapper). It is an optional component provided by this library which gives the data mapper pattern that allows your program to map to the database in multiple ways. If you want to explore more of its uses, you can visit http://www.sqlalchemy.org/

Modules:

SQLAlchemy has a rich set of modules in it, which gives your power to link and access SQL data in a flexible way. Some of its modules are as follows:

Query: Query is the basic source of all the SELECT statements in the SQL database. This method allows you to generate a query for the database.

Add_column: It helps in adding a column expression with the list of query results.

Add_Columns: It helps in adding multiple column expression with the list of query results.

Add_entity: It adds a mapped entry in the list of result.

All: it helps in returning results generated by the query.

As_scalar: It returns the whole SELECT expression given by the query.

Autoflush: It gives a query with particular set of "autoflush"

Column_description: it returns meta-data for the returned query column.

Count: It gives a count of rows from the returned query results.

Delete: it helps in deleting the bulk data from the query results.

www.ingramcontent.com/pod-product-compliance
Lightning Source LLC
Chambersburg PA
CBHW070330220526
45467CB00001B/102